Destiny or Chance: our solar system and its place in the cosmos

Written by a leading planetary scientist, this book tells the remarkable story of how our solar system came into existence. It provides a fast-paced and expert tour of our new understanding of the Earth, its planetary neighbors and other planetary systems.

In a whirl-wind adventure, we are shown how the formation of mighty Jupiter dominated the solar system, why Mars is so small, where comets some from, how rings form around planets, why asteroids exist, and why Pluto isn't a planet at all. *En route* we discover that chance events have shaped the course of the history of our solar system. Dramatic collisions, for example, have caused the tilts and spins of the planets, the extinction of the dinosaurs, and the rise of the human race. Finally, we look at how suitable Earth is for harbouring life, what other planetary systems look like, and whether we are alone in the cosmos.

For all those interested in understanding our solar system and its place in the cosmos, this is a lucid and compelling read.

Stuart Ross Taylor is an Emeritus Professor at the Australian National University with a personal stake in the solar system as the asteroid 5670 was recently named Rosstaylor. His research interests are wide-ranging, including geochemistry, cosmochemistry, origin and evolution of the Moon and planets, meteorites, tektites and the continental crust of the Earth, with side interests in the analytical chemistry of trace elements.

Professor Taylor was a Principal Investigator for the NASA *Apollo* Missions and has written 220 technical papers and six books, including *Lunar Science: A Post-Apollo View*, *The Continental Crust: Its Compostion and Evolution* (with Scott McLennan) and *Solar System Evolution: A New Perspective*. Professor Taylor is a member or honorary member of several scientific academies and is a Foreign Associate of the National Academy of Sciences (USA). He has received numerous awards, including the Goldschmidt Medal of the Geochemical Society and the Leonard Medal of the Meteoritical Society and is a member of the Board of Advisors to the Planetary Society.

The author is married, with three daughters and lives in Canberra, Australia. He is an avid listener to classical music, particular' ~f Haydn and Mozart, and is a keen student of history. Gardening occupies '

STUART ROSS TAYLOR

Destiny
or Chance
our solar system and
its place in the cosmos

'Since one of the most wondrous and noble questions in Nature is
whether there is one world or many, a question that the human
mind desires to understand, it seems desirable for us to enquire
about it'

Albertus Magnus, 1200–1280 AD

CAMBRIDGE
UNIVERSITY PRESS

PUBLISHED BY THE PRESS SYNDICATE OF THE UNIVERSITY OF CAMBRIDGE
The Pitt Building, Trumpington Street, Cambridge, United Kingdom

CAMBRIDGE UNIVERSITY PRESS
The Edinburgh Building, Cambridge CB2 2RU, UK
40 West 20th Street, New York, NY 10011–4211, USA
10 Stamford Road, Oakleigh, VIC 3166, Australia
Ruiz de Alarcón 13, 28014 Madrid, Spain
Dock House, The Waterfront, Cape Town 8001, South Africa

http://www.cambridge.org

First published 1998
First paperback edition 2000

Printed in the United Kingdom at the University Press, Cambridge

Typeset in Monotype Ehrhardt 11/13pt, in QuarkXPress™ [wv]

A catalog record for this book is available from the British Library

Library of Congress Cataloging in Publication data
Taylor, Stuart Ross, 1925—
    Destiny or Chance : our solar system and its place in the cos-
mos /
    Stuart Ross Taylor.
       p. cm.
    Includes index.
    1. Solar system. 2. Planets. 3. Astrophysics. I. Title.
QB501. T248   1998
523.2dc21   9817412 CIP

ISBN 0 521 48178 3 hardback
ISBN 0 521 78521 9 paperback

Contents

Prologue

There are so many books about the exploration of the solar system that has been conducted over the past three decades that one might ask why there is a need for another one. Here I have tried to interpret the new information from this exploration in a broad perspective. Now that we understand the details of our own planetary system, it is fair to raise once again the ancient question as to whether such arrays of planets, satellites, asteroids, comets and the rest are likely to occur around the myriad of other stars. If planetary systems are common, the question then moves to whether copies of our planets exist. Lurking in the background is the expectation that something like the Earth, complete with its set of interesting inhabitants, might exist. My excuse for yet another book on the solar system is to examine this question.

I have written this book in response to numerous suggestions and much encouragement from my friends. It was only after I had made various promises to them that I discovered how easy it is to write about science for specialists and how difficult it is to explain its findings to others, even to scientists outside one's own discipline. Such a course is not without its own perils. One may escape from the jungle of jargon only to find oneself mired in a swamp of clichés.

The book does not follow the usual arrangement of starting with Mercury and marching stolidly out to the giant planets. Instead, the unconventional arrangement I have adopted here has arisen naturally as I have tried to explain how the system came to be and why the various bodies happen to be where they are. Thus, this book is organised into six chapters. The result is that there are many associations that may at first sight seem surprising. Mars finds itself associated with the asteroids, and Mercury with the Moon. I hope that the reasons will become clear to the reader as the story progresses. I have included only a small selection of the excellent and widely available pictures of the planets and satellites to illustrate particular points in the discussion. In a scientific book, it is normal to make reference to the sources of the information that have been used. Here I have tried a different approach, writing more in the style of an essay with only a few formal references.

However, to reassure readers that what I have written is firmly based on fact, this work is derived from my book, *Solar System Evolution: A New Perspective*, which was published by Cambridge University Press in 1992, and which contains over 1150 citations to the scientific literature. Readers who become interested, for example in meteorites, and wish to pursue the subject in more detail, could consult my earlier book. This will then lead into a vast literature on a whole variety of topics, in the case of meteorites with over 240 references, or on the impacts of meteorites, asteroids and comets with over 170 references. I refer people interested in tracking down sources to that work, as well as to the series of books on the solar system published by the University of Arizona Press, of which the following are especially relevant to the theme of this book.

> *Asteroids II* (Editors: R. P. Binzel *et al.*) 1989.
> *Hazards due to Comets and Asteroids* (Editor: T Gehrels), 1994.
> *Mars* (Editors: H. H. Kieffer *et al.*), 1992.
> *Meteorites and the Early Solar System* (Editors: J. F. Kerridge and M. S. Matthews), 1988.
> *Protostars and Planets III* (Editors: E. H. Levy and J. I. Lunine), 1993.
> *Satellites* (Editors: J. A. Burns and M. S. Matthews), 1986.

The sources of various quotations and comments that are identified by numbers in the text (e.g., [8]), are listed by sections in an appendix at the end.

A few technical matters: Time and distance are particularly difficult to deal with in the solar system because both extend far beyond our daily experience. The great contribution of geology to philosophy was to establish the immensity of time. Comments about intervals of time 'as brief as a million years' are common in scientific literature and intervals less than a few million years can barely be resolved in this discussion. I avoid the modern scientific convention that refers to the passage of one billion years as a gigayear (or its even more appalling abbreviation, Ga) because it reduces this stupendous period of time to a trivial level. The origin of the universe dates back around 15 billion years. The time of formation of the solar system, which is known rather precisely, is less than one third of that. Life appeared over three billion years ago on this planet. In contrast, it is only 10 000 years since the last ice age ended and the ice that had covered much of Europe and North America retreated. The whole of recorded civilisation is compressed into the past 6000 years.

Distances in the solar system are usually given in terms of the Astronomical Unit. This is the average distance between the centres of the Sun and the Earth, around 150 million kilometres. This useful unit is abbreviated to AU throughout the text. It should not to be confused with Au, the chemical symbol for gold, nor with Å, the ångström unit, another useful measure, which is about the size of an atom. The planets extend out to the orbit of Neptune at about 30 AU. The outer boundary of the solar system is at the edge of a spherical cloud of comets that extends to about 50 000 AU. Light takes almost a year to traverse this distance to reach us from that distant region. All these immense regions are trivial on an astronomical scale. For these vast distances, the distance travelled by light in a year, the so-called 'light year', that is about 63 000 AU, now becomes a more useful measure. The nearest star is about four light years away.

One of the most striking features of the solar system is that it mostly lies in one plane. This is defined by the orbit of the Earth around the Sun and is also called the plane of the ecliptic. The tilt or obliquity of the planets refers to how far the spin axis of the individual planet is tilted relative to this plane. Thus, the axis of rotation of the Earth is tilted at a little over 24°, a feature that provides us with the seasons that we so greatly admire, as the northern and southern hemispheres receive more or less sunlight.

Two other terms dealing with the orbits of solar system bodies need to be mentioned. These are the inclination and the eccentricity of the orbits. Inclination is the angle that the orbit of the planet, asteroid, comet or whatever makes to the plane in which the Earth rotates around the Sun. Except for Mercury, the planets have inclinations within a few degrees of the plane of the ecliptic. Pluto, here downgraded from planetary status, has an orbit around the Sun that is inclined at 17° to this plane. Another smaller version of Pluto in the outer solar system has an orbit inclined at 24°, while many comets have high inclinations.

How far the orbit departs from a perfect circle and becomes oval or elliptical is measured by its eccentricity. Kepler established that the orbits of the planets are elliptical, although they do not in fact deviate very far from circular. It's different further out. The large Pluto-like object mentioned above has an extremely eccentric orbit that takes it from just beyond Neptune at closest approach to the Sun, out to 130 AU at the farthest point.

Another problem I have to mention deals with temperature scales. In addition to the familiar Centigrade (or Celsius) and Fahrenheit scales, the Kelvin scale is commonly used in science. It uses the same intervals as the Centigrade scale but is expressed simply as K (not to be confused with the same symbol used to indicate 1000) without the usual symbol (°) for degrees. Absolute zero on the Kelvin scale is the temperature at which all motion of molecules ceases. It is 273 degrees below the zero on the Centigrade scale that is set by the freezing point of water. Thus, to convert from Centigrade to Kelvin, one simply adds +273. One of the coldest places in the solar system is the surface of Triton, the satellite of Neptune, which has a surface temperature of a mere 38 K or, on the centigrade scale, −235 °C.

Finally, percentages are the unit that is commonly used in talking about the abundances of the chemical elements. Another convenient unit is 'parts per million', usually abbreviated to 'ppm'. One per cent (one part per hundred) is 10 000 ppm. Ppm is a useful unit for comparing the abundances of trace elements, mainly because it enables us to use small numbers and so avoid long strings of zeros that allow errors to creep in easily. For example, the concentration of uranium in the crust of the Earth is usually referred to as three ppm (rather than 0.0003 per cent), while the total amount of water in the Earth amounts to about 500 ppm, rather than 0.05 per cent.

Parts per billion (or ppb) are employed for abundances 1000 times less than ppm. Thus, the amount of the element iridium in the Earth's crust is only one tenth of a ppb. In contrast, this element is 5000 times more abundant in meteorites, where it is present at 500 ppb or 0.5 ppm. Because of this extreme difference, concentrations of iridium in the crust as low as 10 ppb are 100 times more than the average and so are commonly signatures of the impact of a meteorite on the Earth. The most famous example is that of the asteroid collision that destroyed the dinosaurs. This event left a measurable fingerprint of iridium around the globe from Denmark to New Zealand.

Acknowledgements

Much of this book was written while I was a Visiting Fellow in the Department of Nuclear Physics, in the Research School of Physical Sciences at the Australian National University. Other portions were written while I was a Visiting Professor in the Institute for Geochemistry at the University of Vienna, and when I was a Visiting Scientist at the Max-Planck Institute for Chemistry in Mainz, Germany. I am grateful to all these institutions for their hospitality.

I owe a deep debt to many of my scientific colleagues for advice and encouragement that has extended over many years as I have contemplated the problems of the Earth, the Moon and the solar system and of our place among all these wonders. The list is far too long to include here. It begins with my school teachers of English and concludes with most of the current workers on the problems of the solar system. A special thanks is due to Dr Christian Koeberl of the Institute for Geochemistry at the University of Vienna, who kindly toiled over a first draft of this book and who saved me from various errors. An anonymous reviewer for Cambridge University Press made many useful suggestions, including that of the present title. I am grateful to Clementine Krayshek, who drew the diagrams.

Time line

Time ago

about 15 billion years	Big Bang: apparent formation of the universe
4566 million years	Oldest meteorite age (beginning of the solar system)
about 4500 million years	Formation of the Earth
about 4470 million years	Formation of the Moon
4440 million years	Crust forms on the Moon
about 4000 million years	Appearance of bacterial life on Earth
about 1800 million years	Appearance of complex cells
550 million years	Beginning of Cambrian Period with first extensive preservation of fossils
250 million years	Great extinction of life at Permian–Triassic boundary
65 million years	Cretaceous–Tertiary boundary: extinction of dinosaurs and much else by an asteroid impact
2.5 million years	Beginning of the Pleistocene ice age
about 120 000 years	Emergence of modern *Homo sapiens*
about 25 000 years	Cro-Magnon Man cave paintings
15 000 years	Ice sheets begin to retreat
12 000 years	Domestication of the dog
10 000 years	Beginning of latest interglacial period
4000 BC	Foundation of Sumerian city of Ur
3000 BC	Construction of Great Pyramid at Gisa
1620 BC	Volcanic eruption at Santorini
753 BC	Foundation of Rome

c. 450 BC	Anaxagoras proposes that the Sun is made of iron
c. 430 BC	Democritus suggests that matter is made of atoms
387 BC	Plato founds the Academy near Athens
335 BC	Aristotle founds the Lyceum at Athens
306 BC	Epicurus establishes his school of philosophy
c. 260 BC	Eratosthenes correctly calculates the radius of the Earth
c. 250 BC	Aristarchus of Samos proposes that Sun is the centre of the solar system
Second century BC	Hipparchus produces his star chart
Early first century BC	Lucretius writes *On the Nature of Things*
Second century AD	Ptolemy writes the *Almagest*
391 AD	Temple of Serapis, containing the library at Alexandria is burned
c. 1320	William of Ockham develops his principle of Ockham's razor
1364	Construction of Great Clock at Strasbourg
1543	Copernicus proposes a Sun-centred system
1597	Tycho Brahe leaves Denmark for Prague
1600	Johannes Kepler arrives in Prague
1609	Galileo Galilei makes first astronomical observations with his telescope
1611	A date for origin of the Earth of 4004 BC calculated by Archbishop Ussher appears in the King James Version of the Bible
1637	Rene Descartes publishes his *Discourses on Method*
1682	Edmund Halley views his comet
1687	Isaac Newton publishes the *Principia*
1755	Immanuel Kant publishes *General Natural History and Theory of the Heavens*
1781	The planet Uranus discovered by Herschel

1796	Pierre-Simon, marquis de Laplace writes his *System of the World* and Josef Haydn composes *The Creation*
1801	The first asteroid, Ceres, discovered
1823	Publication of Olbers' paradox (Why is the sky dark at night?)
1846	The planet Neptune discovered
1859	Publication of *The Origin of Species* by Charles Darwin
1929	Edwin Hubble establishes that the universe is expanding
1969	First manned landing on the Moon
1974	Mercury photographed by *Mariner 10* mission
1976	*Viking* missions land on Mars
1977	*Voyager* missions launched to the outer planets
1990	*Magellan* mission begins mapping of Venus
1995	First confirmed discovery of 'planets' around other stars

1

Setting the stage

It has taken us a long time to discover where we are. Primitive tribes living in remote jungle valleys have often been astonished to discover that the Earth extends far beyond their limited horizon and that they are not its sole inhabitants. Before Copernicus, similar views were held in the civilised world. It was generally believed that the Earth was the centre of the universe. However, it was gradually realised that we live in a bigger arena. When you look up at the sky on a dark night in the country, the most striking feature, when the Moon is down, is the glowing band of stars, referred to as the Milky Way, a term first used in English literature by Geoffrey Chaucer (1342–1400) in 1384. This splendid band of stars spreading across the heavens is an edge-on view of our galaxy from the inside.

Although there is a place for the Milky Way in most mythologies, before recent times only Kant seems to have realised what we were looking at. From a nearby galaxy, a few hundred thousand light years away, the magnificent spiral structure that is obscured by our edge-on view would be revealed in all its splendour. But even this enormous spiral system is only a tiny portion of the universe. Each new telescope reveals a larger universe than our imagination had conceived. Like travellers lost in an immense wasteland, we desperately seek for signs that we are not alone. Now we have begun to discover 'planets' circling other nearby stars. Then there is the remote chance that life existed at an early stage on Mars. Both have added new hope to the possibility that we are not alone, lost among the incomprehensible spaces that are extended with each new discovery.

Meanwhile, from the past three decades of space exploration, we have developed a new understanding of the solar system, and of the place of the Earth in it. We understand much about the planets, how they were formed and how they evolved. This enables us to take

another look at the idea of 'one world or many?'. How easy or difficult would it be to make a duplicate of our solar system, or of the
Earth, complete with its interesting cargo of inhabitants. Are habitable planets, complete with 'little green men (or women)' readily
available and common elsewhere? This problem is addressed by looking at what we have discovered about our own system of planets.

I begin by examining what the ancients made of the world in which
they found themselves, as civilisation slowly arose following the melting of the great ice sheets. Most of our present ideas were formulated
in the great flowering of civilisation in Greece and Rome. Following
the collapse of the Roman Empire, there followed a thousand years
of intellectual stagnation in the West. Astronomy survived through
the work of Arab observers; many of our brightest stars such as
Aldebaran still bear their Arabic names. The revival of learning in
Europe led to the Copernican revolution in the sixteenth century.
This created a new world view that the human ego is still trying to
come to grips with.

The place of the solar system in the universe

The view before Copernicus

The comfortable and apparently obvious idea that the Earth is the
centre of the universe no longer attracts much attention. This is
not only because such notions have been replaced by those of the
Copernican Revolution, but because in such models, the origin of the
Earth, Sun and planets was tied to the origin of the universe. After
all, the Earth could hardly be younger than the rest if it occupied the
central position. Now, however, we are aware that the age of the solar
system is only about one third of the age of the observable universe.
This makes it no longer necessary, as was the case with the authors of
the Book of Genesis, to seek a common origin for Earth, Moon, Sun
and stars. Most of this progress has been made by the discovery of
new facts, not by theories. Galileo's observations, like those of Darwin,
have done more to give us a correct view of the world than most of
the thinking about it over the centuries.

The Babylonian and Greek astronomers observed the strange motion of the planets against the fixed positions of the stars. In this manner, they became aware that there were two classes of heavenly objects in addition to the Sun and the Moon. The term 'planet' is derived from the Greek word meaning 'wanderer'. It is curious that although the ancient astronomers devoted much study to the movements of the planets, they did not spend much time considering the origin of the solar system. The planets were mostly not clearly distinguished from the other heavenly bodies. The whole question of origins seems to have been the province not of the astronomers, but of the philosophers. There was no shortage of these, nor of their ideas.

Some astronomers, however, took up the challenge. Among them was Anaxagoras (*c.* 500–428 BC), who considered that the Moon was a stone. He thought that the Sun was a red-hot mass of iron bigger than the Peloponnesus, the southern region of Greece that is about the size of Sicily. This idea that the Sun might be made of iron was based on a reasonable interpretation of the available evidence. An iron meteorite had fallen about 467 BC in ancient Thrace and Anaxagoras concluded that the visitor had come from the Sun. He was banished from Athens because his views about the composition of the Sun and the Moon were considered to be heretical. Little of his work has survived, but apparently he pictured the Earth at the centre of a sort of large cosmic whirlpool. In this he anticipated the notions of Descartes in the sixteenth century, demonstrating the truism that few ideas are truly original.

The great trio of Greek philosophers, Socrates, Plato and Aristotle, whose ideas have formed the basis for western culture, were mostly concerned with questions of purpose. They distinguished carefully between the Earth, with its obvious imperfections, and the heavens, which they held to be unchanging. Four elements, earth, air, fire and water, sufficed to make up the Earth. The heavenly bodies in contrast were composed of shining crystal, a perfect fifth element, or quintessence. The Moon was also made of this. The dark patches that one could easily see on the face of the Moon were thought to be the reflections in this perfect mirror from the mountains and oceans on the Earth.

The doctrine of Socrates (*c.* 470–399 BC) held that the heavens were perfect, in obvious contrast to the Earth. This left no room for any changes or evolution and so did little to encourage scientific investigation. Plato (*c.* 428–347 BC) concerned himself with the motions of the planets rather than their origin. He did suppose, however, that the Earth was moving. In his scheme, the heavenly bodies were supposed to move in perfect circles and the apparently chaotic wandering of the planets among the fixed stars was a major problem. The problem of perfectly circular orbits continued to haunt astronomers as late as Copernicus, over 1000 years later, until Kepler finally broke the spell. Aristotle (384–322 BC), the third member of the trio, also thought that the heavens were permanent and thus not subject to the earthly laws of physics as he perceived them. His views, wedded to the concept of a providential Old Testament God who designed all for our well-being, were to dominate Western culture for 2000 years.

A refreshing contrast to these views was proposed by Aristarchus of Samos, who lived around 250 BC. He placed the Sun at the centre of the solar system, and included the Earth with the rest of the planets. He realised that the Earth was small in relation to the Sun. Many people today have not made that intellectual leap. Aristarchus appears to be the first person who suggested that the Earth both rotates and revolves around the Sun. This idea was not forgotten, but lay around until revived by Copernicus over a millennium later. It is fitting that a prominent crater on the Moon is named for Aristarchus.

Epicurus (341–270 BC), who was a strong critic of the views of Aristotle, did not give the heavens any special or separate status. He supposed that the heavenly bodies formed by random collisions of atoms, whose existence had been proposed by Democritus (about 470–400 BC) 150 years earlier. We would now call Epicurus a materialist. The Epicurean School rejected divine explanations, and believed in physical causes. Unfortunately, it did not encourage investigations into natural phenomena, so that no scientific advances resulted. Epicurean philosophy was mostly concerned with freedom and happiness and was very popular. It survived until the fourth century AD before the Christians managed to defeat it. Our best surviving statement of the physical theory of Epicurus comes from the Roman poet and philosopher Lucretius (96?–55 BC). In his long poem *De rerum natura* (*On the Nature of Things*) he adopted many of the ideas of

Epicurus. He encouraged a materialistic outlook and discouraged superstition. It is refreshing that he paid little attention to astrology, which was popular then as now. What path would the history of the world have taken if the ideas of Epicurus and Lucretius had taken root rather than those of Aristotle?

Among others deserving a special mention, Eratosthenes (276–195? BC) correctly calculated the radius of the Earth. His answer to this classical problem was within about one per cent of the modern value, a technical feat that was not rivalled for the next 1500 years.

Ptolemy is famous for his theory of the solar system. He compiled a summary of Greek astronomical thought and data in his book the *Almagest*. It was a triumph of the use of geometry in understanding the solar system. This work was the definitive work on astronomy until the end of the Middle Ages and so remained the acceptable explanation for over a millennium. Like Lucretius, very little is known of his life, except that he lived in the second century AD. The works of Ptolemy were much studied by the later Arab astronomers. His birth and death dates are unknown, although the Arab sources recorded that he lived for 78 years. Nevertheless, in spite of his great reputation, Ptolemy remains an obscure figure. It is not clear how reliable his measurements were, particularly since he worked for the state religion, which was heavily concerned with astrology. He seems to have been endowed with bad judgment, since he rejected both the Sun-centred solar system of Aristarchus and the essentially correct value for the size of the Earth that Eratosthenes had worked out. Both decisions put the progress of scientific knowledge back for the next 1500 years. Perhaps Ptolemy's major achievement was to salvage the star catalogue of Hipparchus. Hipparchus was the greatest of the ancient observational astronomers and had worked in the second century BC. His catalogue listed 850 stars arranged in six orders of apparent brightness, more or less in line with modern concepts.

Like his Greek predecessors, Ptolemy felt that the imperfect Earth could not be given a place among the heavenly bodies, which were composed of shining crystal in their cosmologies. Echoes of this philosophical approach still appear in the very common tendency to consider unknown or distant regions as uniform in composition. Examples include the deep interior of the Earth, the solar nebula and the universe, all of which were thought until quite modern times to be uniform; more recent information is rapidly dispelling these myths.

The system devised by Ptolemy placed the Earth at the centre of the universe. The motions of the planets followed extremely complicated paths. Despite its theoretical defects, it was a practical success and remained in use up to the late Middle Ages. However, many of its problems had been long understood by skeptical observers. One of these was Alfonso X (The Wise) King of Castille (1221–1284 AD), who is commemorated by having one of the larger craters on the Moon named in his honour.

Laplace, the French scientist who enters the picture later, tells the following story about him.

> Alfonso was one of the first sovereigns who encouraged the revival of astronomy in Europe. This science can reckon but few such zealous protectors; but he was ill seconded by the astronomers whom he had assembled at a considerable expense and the tables which they published did not answer to the great cost they had occasioned. Endowed with a correct judgment, Alfonso was shocked at the confusion of the circles, in which the celestial bodies were supposed to move; he felt that the expedients employed by nature ought to be more simple. 'If the Deity' said he, 'had asked my advice, these things would have been better arranged'.[1]

Despite such opinions, scientific knowledge in Europe by the fourteenth century was less advanced than in Greece and Alexandria in the second and third century BC. The level of mathematics was about that which the Babylonians had achieved two millennia before.

The Copernican revolution

The Copernican revolution is usually dated at 1543. This was the year of the publication of the great work of Nicolaus Copernicus (1473–1543) *De revolutionibus orbium coelestium, libri VI (On the Revolutions of the Celestial Spheres)*. He is reputed to have received the book on the day he died. Few modern authors would care to wait so long.

The model of Ptolemy had placed the Earth at the centre of the universe. This was obvious to everyone and equally agreeable to the ego of *Homo sapiens* (I use throughout the book this scientific term for human beings, thus avoiding the politically incorrect term, mankind, and its ugly politically correct alternative, humankind). After all, it was clear to casual observers that the Earth was flat and that the Sun, Moon, planets and stars all revolved around it. Any child could understand this medieval view of the universe. One is reminded of the current debate over creationism, yet another simplistic view of the world. Furthermore, the Ptolemaic System, for all its complexity, worked well enough for practical matters, including navigation. Columbus used it. Minor problems were accommodated by complicated adjustments until a complex array of epicycles and the like, to which Alfonso had objected, encrusted the whole scheme.

Copernicus however, placed the Sun at the centre. Why did he do this? One can do little more than speculate 400 years later, but he seems to have viewed the Sun-centred system as more intellectually satisfying than the Earth-centred model of Ptolemy. It is curious that Copernicus did not refer to ideas of Aristarchus of Samos, who had proposed a sun-centred system eighteen centuries earlier.

Daniel Boorstin (b. 1914), in *The Discoverers* (1983) records that 'Copernicus possessed an extraordinarily playful mind and a bold imagination'[2] and that his model was driven by aesthetic rather than scientific reasons. But the new idea did not arise in a vacuum, any more than did Darwin's theory of evolution. Along with Alfonso, other thinkers in the Middle Ages, of whom Nicolas of Cusa (1401–1464) and Regiomontanus (1436–1476) were examples, had laid the intellectual framework for dismantling the old system.

The new scheme of Copernicus was not without its problems, and in fact did not work as well as Ptolemy's for practical applications. The planets remained in circular orbits, so Copernicus still had to use even more epicycles than Ptolemy to account for their motions. According to this notion, planets, like a trick cyclist, rotated around in small circles, or epicycles, as they progressed in their circular orbits around the Earth. Epicycles were an obvious solution to the problems of the apparent loops in the motions of the planets as seen from the Earth.

This is seen most easily for Mars, which after moving slowly eastward though the constellations, reverses its normal path and travels westwards, before resuming its slow eastward course among the fixed stars. We now know that this curious reversal that we observe is due to the Earth, with its orbital period of 365 days, overtaking Mars, which takes 687 days to go around the Sun.

It took a long time after the death of Copernicus for the idea that the Earth goes around the Sun to be commonly accepted. In our age, Darwinian evolution is likewise taking some time to become established as the accepted world view. The next significant step in understanding the solar system was taken by Tycho Brahe (1546–1601), another outstanding figure of Renaissance science. His chief accomplishment was the precise measurement of planetary positions. This was carried out by naked-eye visual observation, as the telescope had not yet been invented. His observatory was on the island of Hven, a short sail from Copenhagen. He was also concerned about the problems with the complicated system of Ptolemy. So he produced a model in which the Sun and the Moon indeed went around the Earth, as everyone could see. However, he had the other planets rotate around the Sun. In this way, he had a foot in both camps. This compromise cosmology was popular, since it appealed to common sense observations and did not conflict with the scriptures. Variations survived until late in the seventeenth century, finally vanishing as the motions of the planets became well understood.

Tycho had other problems. He lost part of his nose in a duel and wore one made of tin for cosmetic reasons. He also disgraced himself in the eyes of his aristocratic family by marrying a peasant's daughter. Finally, he was so unpopular with the other residents of his island that they demolished his observatory when he lost royal favour and he had to move with his data to Prague in 1597.

Here, chance played its role. Just in time, another refugee arrived in Prague in 1600. Johannes Kepler (1571–1630) had been banished from the pleasant town of Graz in Austria, a victim of Catholic persecution. He became Tycho's assistant and succeeded him as Imperial Mathematician when Tycho died suddenly in 1601. Kepler thus inherited or perhaps just took ('usurped' was his word) the boxes that contained Tycho's monumental observations. These data formed the basis for Kepler's basic discoveries of the laws of planetary motion. Kepler's great contribution was to get rid of the notion that had survived since Aristotle that the planetary orbits were circular.

He discovered that the orbits were elliptical and became an advocate of the Copernican System.

However, like many other scientists, he was mainly concerned with other matters so that, as one author has commented, 'the three major gems in Kepler's works on astronomy lay in a vast field of errors, of irrelevant data, of mystical fantasies, and of useless speculations'.[3] It is difficult to imagine the intellectual climate in which he lived. His mother was accused of witchcraft and he spent several years defending her, ultimately successfully, from the appalling fate that accompanied conviction.

Despite such distractions and with a vast amount of labour, Kepler was able to fit the orbits of the planets into spheres based on the five 'perfect' geometrical solids; cube, tetrahedron, octahedron, icosahedron and dodecahedron. These are the only solids bounded by identical faces and were so considered 'perfect'. They have long fascinated philosophers. Plato had used the first four forms as the basic shapes for earth, air, fire and water, while the dodecahedron was the model for the heavens.

Kepler considered that he had answered a fundamental question; why were there only six planets, with five intervals between them (as known at the time)? Kepler's view was that this cosmic limit was imposed because of the small number of 'perfect' solid forms. However, the planetary orbits, on the basis of Kepler's own laws, turned out to be elliptical, not circular. Thus, his elaborate geometrical system fell into ruin.

Clocks had been prominent features in town squares in Europe since the fourteenth century. They became more sophisticated as clockwork became perfected and often included astronomical models as well as religious displays. One of the earliest was constructed by Richard of Wallingford in 1320 at St Albans, in England during the reign of Edward III. Another famous example is the great clock at Strasbourg, dating from 1364. Others were at Mantua, Padua, Prague and Venice, Such mechanical marvels led to the idea that perhaps the universe was some kind of giant clockwork. A clock requires a builder, suggesting that the universe had been created by a master craftsman.

Once the solar system had been constructed by an omnipotent clockmaker and the system was set running, no further attention was needed. It would continue to operate under the laws of physics. Such ideas went back to Nicolas of Oresme (1330?–1382), a bishop who had conceived of God as the master clockmaker. Kepler was an enthusiastic supporter, suggesting that perhaps magnetism was the driving force, just as falling weights drove earthly clocks.

The clockwork idea was also consistent with the Bible. An Irishman, Archbishop Ussher (1581–1656), calculated that the creation of the world (including the universe) had occurred in 4004 BC on Sunday, October 23 at 9.00 am. This date, although now derided, was carefully derived from the available biblical record. What it represents is essentially that of recorded history. The earliest city, Ur, was founded around that time. It was generally accepted at the time, even now appearing in many editions of the Bible. The significance of this date, if correct, was that the universe had not had much time to evolve and everything must have been created in the beginning, more or less as it appeared now.

The Copernican revolution did not resemble those of more modern times. Fifty years after the publication of his system by Copernicus, little had changed. His ideas had disturbed neither the public nor the church. What was needed was some crucial observation to decide between Copernicus and Ptolemy. This came, as is usual in scientific progress, with a technical advance. The telescope had been invented about 1600 by Hans Lippershey, a Dutch spectacle-maker, apparently by accident. When the news reached Italy, the Senate of Venice asked Galilei Galileo (1564–1642), a skilled maker of instruments, to make some. He was the son of a lute player and composer, but had decided not to follow his father's career. We are still living with the consequences of that decision. It was not of course the intent of the Venetian state to upset the accepted view of the world. Their reasons were more down to earth. Telescopes would obviously be useful for an empire based on sea power. One is reminded that the British Admiralty did not send out HMS *Beagle*, carrying Charles Darwin, because they wished to change our view of nature or overturn the authority of the scriptures. They wanted better charts of the South American coast.

Galileo's observations are famous. The Milky Way was composed of stars, and so maybe the universe was infinite. The Moon was not a smooth mirror after all, but rough like the Earth and so perhaps made of the same material. Venus showed phases like the Moon, including a full face. This told Galileo that Venus was passing behind the Sun. Another critical observation that led to the collapse of the Ptolemaic System came when Galileo discovered in 1610 that four satellites were rotating around Jupiter.

Copernicus was right after all. The idea that the Sun, rather than the Earth, was at the centre of the universe caused a profound change in the view of our place in the world. It created the philosophical climate in which we live. It is not clear that everyone has come to grips with the idea, for we still cherish the idea that we are special and that the entire universe was designed for us.

Rene Descartes (1596–1650) then took up the challenge of the origin of the solar system. His view of the world was a completely mechanical one. He postulated that there was no basic difference between the forces driving a clock, the solar system and living matter. He proposed that the universe contained many circular eddies. Like a whirlpool, matter accumulated in the centre of the vortex to form the Sun. Coarser particles were captured to form the planets. Satellites formed in secondary whirlpools surrounding the planets. He appears to have disregarded some of the conventions of his age if John Aubrey's (1626–1697) biographical sketch is a realistic account. Amongst other gossip in Aubrey's *Brief Lives* is the statement that Descartes 'was too wise a man to encumber himself with a wife, but he kept a handsome woman, by whom he had some children'.[4]

By the time that Isaac Newton (1647–1727) appeared, the Copernican System had long dominated thought. Newton's work was the culmination of the work of Copernicus, Kepler and Galileo. Writing in 1704, Newton was impressed by the tidy nature of the solar system. He was irritated by the qualitative notions of Descartes, and showed that the complexity of the solar system could be dealt with by exact physical laws. The planets were securely tucked into their orbits and the space between was apparently clean. Newton assumed that the world had been created essentially in its present form only a few thousand years before, according to the biblical time scale that Archbishop Ussher had calculated. This left no time for the system to evolve from a more primitive state as Descartes had imagined. Thus, it required a Creator, who had ordered each planet to move in its particular orbit.

The success of Newtonian mechanics reinforced the notion that the solar system was some type of celestial clockwork. This theme of

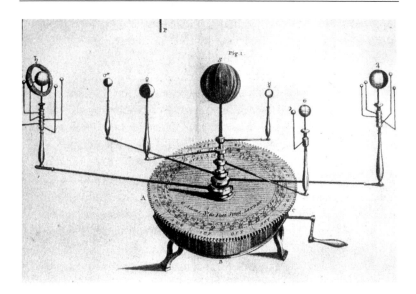

Figure 1. A simple orrery from the late eighteenth century.

a celestial clockmaker came to dominate thinking about the solar system in the seventeenth and eighteenth centuries. These ideas bore fruit in the construction of mechanical models of the solar system. In fact, models of the solar system date back to antiquity. Cicero (106–43 BC) in *De republica* tells of having seen a model that Archimedes (287–212 BC) had built. It showed the Sun, Moon and the five planets known to the Ancients.

The eighteenth century models were named orreries after the 4th Earl of Orrery, Charles Boyle (1676–1731). These instruments (Figure 1) became very popular. There is a beautiful example in the Meteorite Hall of the Natural History Museum in Vienna of a 'Kopernikanische Planetenmaschine' made in 1761 for the Austrian Emperor. When Louis XV (1710–1774) constructed a new wing at Versailles, an orrery was placed in the central room, in contrast to the chapel which forms the centre of the old wing. This was in keeping with the philosophy of the Age of Enlightenment.

However, Newton noted that there were small variations in planetary orbits, so in his system God had to intervene from time to time to make periodic repairs or adjustments, in effect winding up the clockwork. This led to complaints by his great rival, Leibnitz (1646–1716), that Newton was guilty of heresy by supposing that God had created something less than perfect. Given supreme power, the construction of a well ordered planetary system should not be beyond the powers of a competent clockmaker. Surely God would not have constructed an imperfect system and would have had enough foresight to create perpetual motion, rather than acting as a maintenance man, who had to wind up the clock and make fine adjustments to the planetary orbits.

A little later, the great philosopher Immanuel Kant (1724–1804) considered the problems of the solar system. In his time, philosophers worked on significant problems rather than on intellectual curiosities. He provided a correct explanation for the Milky Way, proposing that it was an edge-on view of a disk of stars. His suggestion that the fuzzy lentil shaped nebulae were distant island universes similar to the Milky Way showed remarkable foresight. This was a leap in understanding that was not confirmed until the third decade of this century, nearly 200 years later. These essentially correct insights perhaps explain why he gets so much credit for his ideas about the origin of the solar system. He felt that the solar system could not arise purely from the mechanical ideas of Newton, but that God had to guide the design. Once this perfect plan was set up, the universe had no freedom to deviate from it.

Kant's model for the origin of the solar system was based heavily on an analogy with the galaxies. It began with a chaotic distribution of particles. This material was assumed to be rotating and to develop into flattened rotating disks. The Sun formed at the centre, and the planets formed at secondary condensations within the disk. He postulated the existence of many additional planets outside the orbit of Saturn, with a gradual transition to the comets. In his book, *General Natural History and Theory of the Heavens* (*Allgemeine Naturgeschichte und Theorie des Himmels*), he populated all the planets with intelligent creatures. They became more clever with distance from the Sun. Thus, a monkey on Saturn would be smarter than Newton.

When his ideas on the origin of the solar system are examined more critically, they turn out to be mostly vague statements. The many contradictions in Kant's hypothesis do not agree with the popular acclaim that it has received. Perhaps this is due to his eminence as a philosopher. It shows how difficult it is to account for the solar system when one of the foremost thinkers of the Enlightenment failed to produce an acceptable explanation. His model is often linked incorrectly with that of Laplace, to which we now turn.

Laplace and his followers

We can date modern thinking about the origin of the solar system from the appearance in 1796 of the *System of the World* by Pierre-Simon, marquis de Laplace (1749–1827) (Figure 2). His work on celestial mechanics, although less well known in the English-speaking world, rivals that of Newton. Laplace was impressed, as Newton had been earlier, with the regularities in the solar system as it was known in the late eighteenth century. The planets all lay in a plane, and they all moved in the same anticlockwise direction around the Sun. The satellites revolved around their parent planets in the same direction

Figure 2. Pierre Simon, marquis de Laplace (1749–1827).

(Laplace ignored the inconvenient fact that at least two satellites of Uranus, discovered by Herschel (1738–1822) in 1787 were orbiting in a plane perpendicular to the rest of the solar system). The orbits of the planets, although elliptical as every schoolchild now is told, are in fact nearly circular. This regular arrangement led Laplace to the concept that the system had arisen far in the past from a primitive rotating cloud, the 'solar nebula'. This idea has survived. This was in contrast to the ideas of Newton, who had believed that the solar system had been created in its present form only a few thousand years earlier.

Laplace however was an inhabitant of the Age of Enlightenment. Born into what we would now call a middle-class farming family, he had survived the French Revolution and was a distinguished member of the French scientific establishment at the beginning of the nineteenth century. He was able to show that the apparent variations in the orbits of the planets were self-correcting and so God was not needed to adjust the system. Laplace gave a copy of his famous book to Napoleon, to whom he had taught mathematics when the Emperor had been an artillery cadet. Bonaparte, seeing no mention of God, presumably the designer of the system, asked Laplace about this omission. Laplace, having solved the problem that had bothered Newton, made his famous reply that he had 'no need for that hypothesis'.[5]

A watershed had been crossed. Now the solar system could be considered as having arisen by the operation of natural processes from a primitive beginning, rather than being created perfect in the instant. This marks the beginning of modern attempts to understand how the Sun and the planets came into being.

At the same time that Laplace was writing the *System of the World*, Josef Haydn (1732–1809) was composing *The Creation*, which he also finished in 1796. It was first performed in Vienna in April, 1798. This oratorio, for five soloists, choir and orchestra, takes over two hours to perform. It is the finest musical statement on the origin of the solar system. It is arguably Haydn's greatest work and it must be considered one of the major triumphs of western civilisation. Haydn's sources were *Paradise Lost* published in 1667 by John Milton (1608–1674) and the biblical account in Genesis. So Haydn had to form the Earth, its inhabitants both animal and human, and the heavenly firmament within the seven days allotted by the authors of the Book of Genesis. More recent work has relaxed this tight time frame, so that 15 billion years or so are now available in which to reach our present position. We are still waiting for artistic statements of the stature of *The Creation* that incorporate our new understanding.

The extent of the universe

In order to obtain some perspective on the place of the solar system in the scheme of things, it is useful to contemplate the scale of the universe as we perceive it at present. It is also useful to bear in mind the dimensions of our own system as new planetary systems are found (see Figures 3 and 4).

The average distance between the centres of the Sun and the Earth is one Astronomical Unit, or around 150 million kilometres. This useful unit is abbreviated to AU throughout the text. There is some more information about it in the Prologue.

The diameter of the Sun is about one hundredth of an AU. Mercury is close to the Sun at 0.4 AU, Venus lies a little inside the orbit of the Earth at 0.7 AU while Mars is just beyond at 1.5 AU. The asteroids fill the gap between Mars and Jupiter, and are mostly concentrated between 2 to 4 AU. The spacing of the giant planets can be easily remembered: Jupiter is close to 5 AU, Saturn is at 10, Uranus at 20, while the planetary system extends out to the orbit of Neptune at

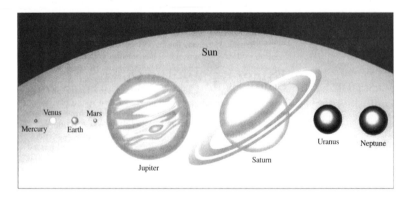

Figure 3. Relative sizes of the Sun and the eight planets.

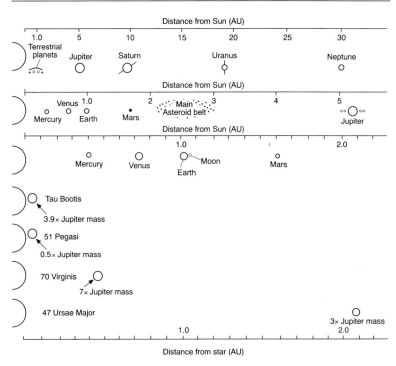

Figure 4. The distances of planets from the Sun in our solar system (top) and distances of planets in other planetary systems from the parent star (bottom) are shown to scale. The sizes of the planets are exaggerated. The top row shows the Sun and planets out to Neptune; the second row shows the solar system out to Jupiter and includes the asteroid belt; and the third row shows the terrestrial planets, Mercury, Venus, Earth and Mars. Some of the 'new planets' around other stars are shown below to the same scale. They are distinctive in that planets larger than Jupiter orbit very close to the parent star.

about 30 AU. Not far beyond Neptune, extending out to over a thousand AU lies a rather flat cloud of icy comets, of which Pluto is probably an escaped member. Further out lies a more spherical cloud of comets, an array totalling perhaps over a trillion in number. This great swarm extends to about 50 000 AU and marks the outer bound of the solar system. This is so far away that light takes nearly a year to reach us from that distant region.

The positions of the planets as one goes away from the Sun, follow a simple mathematical rule. The distances of the planets from the Sun can be expressed as a series: 0.4, 0.7, 1.0, 1.6, 2.8, 5.2, etc., which is close to their separation in AU. This sequence of numbers is arrived at by adding a constant 0.4 to the doubling sequence of 0, 0.3, 0.6, 1.2, 2.4, 4.8, etc. This interesting relationship was first discovered by Johann Daniel Titius von Wittenberg (1729–1796). It was later drawn to popular attention by Johann Elert Bode (1747–1826), and so is correctly termed the Titius–Bode Rule.

This regular spacing of the planets has always attracted wide interest. It is usually cited as one of the significant features to be taken into consideration in any theory for the origin of the solar system. It became famous when the asteroid Ceres was discovered in place of the 'missing planet' at 2.8 AU between Mars (at 1.6) and Jupiter (at 5.2).

However, the relationship is only approximate. It works well enough out to Uranus, but breaks down at Neptune. This planet should be out at 38.8 on Bode's scale, but it resides, inconveniently for the rule, much closer in at 30 AU. The position of Pluto does not fit either, but this left-over icy body will be excluded in due course since, as I will explain later, it does not justify classification as a planet.

Does the Titius–Bode rule have any real significance? It seems reasonable to expect that if the rule represents some major physical factor in building planetary systems, then some other major properties might also vary with the simple mathematical regularities in spacing of the planets. However, it is curious that there is no correlation with mass or composition, either with the spacing given by the rule, or with distance from the Sun. This raises the possibility that the rule is a secondary, rather than a primary, property of the solar system.

There is, of course, no real evidence that we are looking at the initial orbital spacing of the planets. What seems likely is that the spacings between the planets have arisen naturally by tidal forces after they were formed. Thus, the famous rule appears to be without much significance for the origin of the system.

Certainly, from the fragmentary evidence available from the newly discovered 'planetary' systems around other stars, the 'rule' does not appear to operate in those locations either. The planetary spacing is different from our own. Some 'planets' larger than Jupiter reside much closer to their star than the orbit of Mercury and orbit their parent star every few days. It seems clear that the famous rule is not some kind of universal rule of thumb for making planetary systems. It's just the way that our planets adjusted to the tidal forces. Other systems have different spacings.

When we look out beyond the solar system into the universe, all the immense distances between the planets become trivial on a galactic scale. It is time to change units. The distance travelled by light in a year, about 63 000 AU, now becomes a more useful measure.

The nearest star is Proxima Centauri. It is the faintest member of a triple star system of which Alpha Centauri is the brightest. This star is familiar to dwellers in the southern hemisphere, as it forms one of the Pointers to the Southern Cross. Light from this nearest star takes over four years to reach us. Although Proxima Centauri is the nearest star at present, the dwarf star Ross 248 will succeed to the title in about 33 000 Earth years. Due to the slow relative movements of the stars, our familiar constellations, such as Orion the Hunter and his companion, the Great Dog (Canis Major), will be rearranged and replaced by other groupings in the future. Edmund Halley (1656–1742), of comet fame, seems to have been one of the first to have realised this, by observing that the positions of many stars in the early eigtheenth century differed from those recorded in the catalogue of Hipparchus in the second century BC.

The Milky Way galaxy is about 85 000 light years in diameter and is rotating slowly. The solar system and ourselves reside in one of the dusty and gas-rich spiral arms (the Sagittarius Arm) about 25 000 light years from the centre. The galaxy turns in a slow majestic wheeling motion. It has rotated less than 20 times since the solar system formed, as it takes about 250 million years to make one revolution. The grand scale of this movement can be appreciated only on geological time scales. Two hundred and fifty million years ago, the Paleozoic Era was drawing to a close, a time marked by one of the great extinctions, when over 95 per cent of life on Earth was extinguished, including the trilobites that had existed for 300 million years.

The nearest major galaxy, M31 or Andromeda, is two million light years distant, and forms one of at least 25 members of the Local Group of galaxies. These include the Magellanic Clouds that are clearly visible in the southern hemisphere. They were seen by and are

named for the Portuguese navigator, Ferdinand Magellan (1480?–
1521). He commanded, but unfortunately did not survive, the first
expedition to circumnavigate the Earth. Beyond the Local Group
extends an apparently endless array of galaxies. The Hubble Space
Telescope reveals that there are perhaps 500 billion galaxies in the
observable universe, an increase by a factor of five over previous
estimates.

Galaxies

These are the most obvious major components of the universe and
need to be mentioned here, as the solar system resides within one.
Typically they each contain 100 billion stars. Although there are a
great multitude of galaxies, it is surprising that the nearby ones at least
fall into only a few general categories (ellipticals, spirals and dwarfs)
as they progressively lose their bulges and transform into more aes-
thetically pleasing disks. The standard model for the formation of a
spiral galaxy (see Figure 5), including our own, begins with a spherical mass of gas from which stars begin to form. The sphere collapses
to a rotating disk within a few hundred million years, leaving a halo
of globular clusters of stars to outline its original extent. I will describe
more about these old halo stars shortly. In this model, the evolution
of the Milky Way galaxy began with the formation of the halo at
around 12 to 15 billion years ago. As the galaxy collapsed to a disk,
star formation began in the spiral arms around ten billion years ago.

It now appears that galaxies may have a much more complex history than previously imagined. Like the terrestrial continents, they
seem to be composed of many separate units that have been swept
together. Thus, galaxies probably do not evolve in isolation, but may
have undergone many collisions.

It seems worth recording an example of a very distant galaxy. It has
a pedestrian designation (6CO140+326). This very distant galaxy,
although it might have been expected to be youthful, seems to be
unremarkable, with an apparently old population of stars. There are

Figure 5. (*a*) A typical spiral galaxy, NGC 2667 (courtesy Anglo–Australian
Telescope). (*b*) Our galaxy. A sketch of the probable structure of the Milky

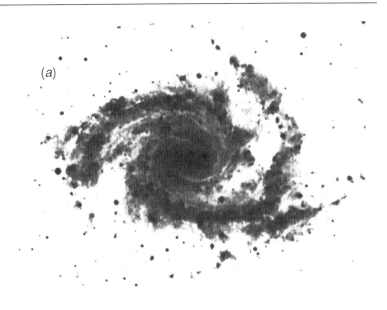

(a)

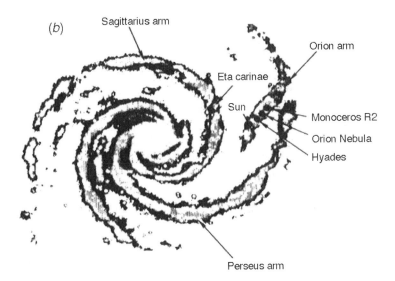

(b)

Sagittarius arm

Orion arm

Eta carinae

Sun

Monoceros R2

Orion Nebula

Hyades

Perseus arm

Way galaxy, showing three nearby arms, and the positions of regions of new star formation and of the Sun.

many other remote galaxies that also appear very similar to those near us. As we see further back in time, the universe seems to get both larger and older than we previously imagined it to be. As this trend has persisted over the entire history of looking at the heavens, no doubt it will continue.

Although it was previously thought that most galaxies formed very early in the history of the universe, it is now clear that galaxy formation has been a continuing process. So the splendid galaxies that we admire are not permanent, but evolve with time, like everything else in nature. Like many objects, they would not have been predicted if they hadn't been observed. How all these beautiful structures arose from the primordial soup of fundamental particles is one of the major questions of cosmology. In the inflationary scenarios that I discuss shortly, they perhaps arise from minor fluctuations as the Big Bang fireball expands.

Is the universe uniform?

I need to address this question briefly, again to place our solar system in a cosmic perspective. A major shift in our understanding of the structure of the universe has occurred within the past few years. Before about 1980, it was generally accepted that the universe was uniform, with galaxies distributed evenly out to the limits of vision. This view has changed dramatically and the structure is now known to be far from random (see Figure 6). Galaxies are distributed along chains, sheets, filaments and in knots. The largest sheet-like structure that has been observed is the 'Great Wall,' which contains many thousands of galaxies and is over 500 million light years long.

In many cases, galaxies appear to be located on the surfaces of spherical shells surrounding dark regions apparently devoid of galaxies. Such structures have been compared to soap bubbles. Two dimensional pictures of galactic distribution look like filaments apparently because the galaxies congregate around the edges of the very large empty bubbles. These great bubbles are about 150 million light years across, and apparently are empty. The universe seems to be forming itself into larger and larger units as it gets older, the same trend that empires on Earth have followed, to their ultimate ruin. Clusters of galaxies appear to have formed within the last few billion years and

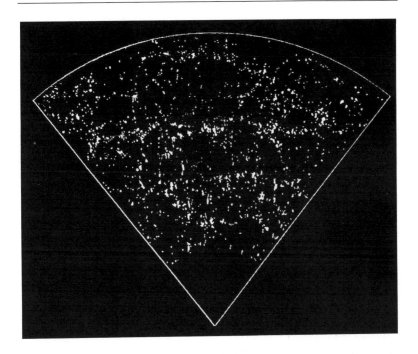

one billion
light years

Figure 6. Galaxies are not distributed uniformly in the universe but occur in clumps and strings. This shows the lumpy distribution of galaxies out to about two billion light years, based on a survey of 24000 galaxies (Carnegie Institution Las Campanas Survey).

these clusters are now beginning to be bound by gravity into super-clusters.

The expansion of the universe

The discovery in 1929 by Edwin Hubble (1889–1953) that the universe was expanding is now known to all. Whether the universe is open and will expand forever, or is closed and so will collapse in the future, depends on the density of the universe.

The universe appears to be extremely flat out to the current horizon around 15 billion years. Thus, it seems to be just balanced between expanding indefinitely and collapsing back. However, when we add up all visible material, it only amounts to around ten per cent of the density needed to cause the universe ultimately to collapse back. For that to happen, ten times as much material is needed. Cosmologists use the term 'omega' for this critical density and it needs to be precisely equal to 1.0. Theory thus demands that 90 per cent of the universe is present as 'dark matter' that we can neither see, nor currently detect. Factors of ten should count for something, even in astronomy, as John Updike once remarked.

What is the observed density of the universe? The values are provocative. The amount of matter that is visible, stars, galaxies and such, don't amount to much. Planets contribute only a trivial amount, although it is important to us since we are standing on some of it. Part of the dark matter may consist of the burnt-out remains of massive stars formed at an early stage. Other candidates are very faint stars, brown dwarfs and such creatures that I talk about later, but there seems to be a scarcity of them. All these are lumped together as 'massive astronomical compact halo objects', or MACHOs, another example of the acronyms that have come to plague us, as an apparently enduring legacy of the Second World War.

I will come back to this problem of the density of the universe when I discuss its origin shortly. Meanwhile, I need to consider how old it is.

The age of the universe

The reason for raising the question of the age of the universe in a book on the solar system is to clarify the difference between the two. In the past, the Sun, stars and planets have often been linked together in attempts to explain their origin. Few creation stories or religious explanations make any distinction.

The age of the universe, that is, the time since the Big Bang, is in dispute with estimates ranging from around 12 to 18 billion years. We can set some lower limits. There are several pieces of evidence that indicate that the universe has to be older than ten billion years. The first is the definitive evidence that the observed abundances of the elements and isotopes of long-lived elements, such as uranium and thorium, which are continuously produced in red giants and supernovae, require well over ten billion years to reach their presently observed values.

The ages of the old globular star clusters that surround the Milky Way galaxy have usually been estimated at 14–16 billion years. Data from the *Hipparchos* satellite (a play on 'Hipparchus', who made the ancient star catalogue) indicate that they may be younger, so that they may match the young ages derived from current measurements of the expansion rate that I talk about shortly. A period of a few billion years is required before these clusters of stars arise to allow for element and star formation following the Big Bang.

It has also proved possible to measure the ages directly of some of the old stars in the halo of our galaxy. This is an entirely independent measurement that does not depend on theories about the nature or expansion of the universe. The age is established by measuring the abundance of thorium in the star. This radioactive element takes about 14 billion years for half its atoms to transmute to lead. Through this radioactive decay, we are provided with a clock that tells us that some stars in the halo of the Milky Way are about four times older than the Sun, whose formation we date quite precisely at four and a half billion years ago. These halo stars thus may have ages of the order of 17 or 18 billion years, with uncertainties in the estimates of two or three billion years.

Finally, it is worth noting once again the presence of very distant galaxies that appear very similar to those close by. Since the stars in these galaxies take time to evolve, ages of the order of 15 billion years are implied. Thus, it is difficult to suggest that the universe as we see it is much less than about 15 billion years old.

The point of all this discussion in our context is that the solar system has a well established age of four and a half billion years (4566 million years to be precise). So whatever age of the universe is finally agreed upon, it is a firm conclusion that the solar system is much younger. It formed late in the history of the universe and its origin is clearly disconnected from it. This is in contrast to most creation accounts. Thus, in the story in the Book of Genesis, the Earth appears first, followed on the fourth day by the Sun, Moon and lastly by stars. Here we start with the stars. The fundamental philosophical conclusion is that our system of planets arose by normal physical processes in a much wider system, as Laplace had perceived. Galaxies had been forming and stars had lived and died for several generations before the solar system condensed in a spiral arm of the Milky Way.

The time from the 'origin' of the universe in the Big Bang can be calculated by measuring the rate of recession of the galaxies. This measure is known as the Hubble 'constant'. Much confusion has arisen over its value. After 70 years of intense effort, an agreed value has yet to be arrived at, with values ranging between about 40 and 90 (expressed as kilometres per second per megaparsec: a megaparsec is 3.26 million light years). The lower the value, the older the universe becomes. A value of 40 implies an age around 25 billion years, while the universe is only 8 or 9 billion years old if the Hubble constant is around 90.

These young ages conflict with the other evidence that the universe is perhaps 15 billion years old. Living in a universe that is younger than some of its parts clearly presents some difficulties. Although the astronomical community has long been divided into these high and low camps, the values for the Hubble 'constant' are coming a little closer, with estimates around 50–60 and 70–80 now being championed by the two groups. Perhaps a resolution is in sight for this previously intractable problem. Astronomers are still measuring the Hubble constant over tiny distances compared with the extent of the universe. Measurements at much greater distances may eventually yield the 'true' value, or give us a better understanding of the density of the universe.

Clearly, we should keep open the possibility that we are still observing a small corner of the universe. Our cosmic horizons have continually expanded with each advance in technology and telescopes, and further significant advances in cosmology may be expected.

How did the universe begin?

It is not the place of a planetary scientist, accustomed to dealing with rocks, to enquire too deeply into this matter, but again it helps to provide a little perspective on our place in the cosmos. At any popular lecture on the solar system, one can guarantee that there will be two questions, the first about UFOs[6] and the second about the Big Bang. Here, I defer to the excellent account of the Big Bang that is to be found in *The First Three Minutes* (1977) by Steven Weinberg (b. 1933).

A period of rapid inflation at an early stage of the Big Bang has been postulated to overcome the problems of expanding the universe from a very dense state. This model predicted that tiny irregularities during this inflationary expansion could account for the observed lumpy nature of the universe and so become the seeds that eventually grow into galaxies. Such very small variations have indeed been found in the radiation that represents the faint residual glow from the Big Bang, thus reinforcing the inflationary hypothesis.

However, a consequence of this model is that the overall density of the universe must be close to the value needed to ultimately stop the expansion. The famous term 'omega' must be equal to one. The problem, as I mentioned earlier, is that the amount of material that we can observe or infer in the universe amounts to not much more than one tenth of that needed. Nine tenths is missing somewhere. To us, empty space appears to be of low density, but the theory tells us that it was ten times denser in the early universe.

Cosmologists are equal to this challenge and have proposed various nuclear particles as the candidate for the missing dark matter. WIMPS, or 'weakly interacting massive particles', are favourites and an equally massive search is underway to detect these exotic and elusive creatures, beside which leprechauns might seem to be firmly rooted in reality. However, we have been surprised before. The Earth is round, not flat and revolves around the Sun. Both notions appeared to be ludicrous propositions to primitive, and not so primitive people. But perhaps what we see is what there is. If the universe were as dense as required by theory, the expansion would have been slowed by gravity over time. This does not seem to have happened. Astronomers looking back to over seven billion years ago (perhaps half the age of the universe) see little evidence for the predicted slowdown. So the universe seems to have too low a density to halt the expansion. Will it fly apart for ever, or does the theory need some revision? It's truly an interesting time for cosmologists.

Three pieces of evidence are generally used to support the presently accepted Big Bang model. One is that a faint glow, the embers from the high temperatures of the Big Bang, is diffused through space. Due to the expansion this apparent remnant from the Big Bang is now observed red-shifted to wavelengths of around one centimetre, in the microwave region. The temperature has dropped to 2.73 Kelvin, close to absolute zero. As this is the first mention of the Kelvin temperature scale, readers unfamiliar with it may wish to consult the Prologue for more information.

The second piece of evidence in favour of the Big Bang is that the abundances of deuterium, helium and lithium that we observe have been held to agree with the predictions of that theory. The third piece of evidence is the entertaining observation that the sky is dark at night.

The darkness of the night sky

This turns out to be one of the more interesting questions in cosmology. Why is the sky dark at night? Corin, the simple shepherd in Shakespeare's *As You Like It* knew that 'a great cause of the night is lack of the Sun',[7] but the question is a little more complex. If the universe is infinite, and filled with stars, then every line of sight must eventually intercept a star. Accordingly, the night sky, and the daylight sky for that matter, should be ablaze with stars.

The problem had been around for a long time. Thomas Digges (1546?–1595), writing in England in 1576, raised the question. He believed that the universe was infinite and that absorption of light from distant stars was responsible for the dark night sky. Kepler thought about the problem, deciding that it showed that the universe could not be infinite. What we were looking at between the stars was the darkness outside the universe. But the problem only became famous as 'Olbers' Paradox' when the German astronomer, Heinrich Olbers (1758–1840), publicised it in 1823. We are fortunate that Edward Harrison (b. 1919) has written an elegant synthesis, *Darkness at Night, a Riddle of the Universe* (1987), giving the many ingenious solutions proposed for this problem.

The fact that the sky is dark at night shows that the universe is not infinite, nor eternal, nor filled with stars. There are several reasons. Stars have restricted lifetimes and burn out over periods ranging from millions to billions of years. The expansion of the universe since the Big Bang has spread out the galaxies and stars. Light from more distant stars and galaxies has been shifted to longer wavelengths outside the visible range. If the universe as we see it is 15 billion years old, then light from more distant regions has not had time to reach us. At the beginning, the sky would have been brilliant. Now the Big Bang has faded to a faint glow just a little under three degrees above absolute zero.

Early insights to the correct solution were proposed by Lord Kelvin and curiously enough by the author Edgar Allan Poe (1809–1849) in his poem *Eureka*. He realised that as we look out, we look into the darkness that existed before the universe. Clearly, one should listen to the poets.

However, it would be premature to suppose that we have an ultimate solution to the fundamental problem of the origin of the universe. There are a few problems with the standard Big Bang model. Although the observed abundances of the light elements constitute a widely accepted 'proof', this has been seriously questioned. The problem depends heavily on the value assumed for the amount of helium produced originally. Some recent estimates place it below that predicted to form in the Big Bang. Finally, the existence of normal looking galaxies at great distances and the old ages obtained from the thorium abundances in ancient stars appear to this reviewer evidence serious enough to suggest an older, rather than a younger, universe.

The notion that the universe began at some definable point has always been philosophically unsatisfactory. The Big Bang has been called an event without a cause. However, despite such problems, it represents the only currently acceptable scientific explanation for the origin of the universe. The question is open, like much else in cosmology. What is needed are new data, not more theories. Now it is high time for this student of the planets to return to his more familiar neighbourhood.

Stars and the Sun

A common or garden star

The Sun and the solar system (and ourselves) are latecomers in the universe. The universe was in existence for something over ten billion years or so before the formation of the solar system. Another four and a half billion years passed before *Homo sapiens* arose to survey the surroundings. When the solar system formed, the universe had long settled down into its present familiar appearance, complete with galaxies and stars and would look much the same as today. The chemical elements made in earlier stars had been dispersed out into interstellar space as those stars died. These events had been going on for countless ages. Like the Mills of God that grind very slowly, these processes had converted only about two per cent of the primitive hydrogen and helium into heavier elements over this immense period of time. The elements included carbon, oxygen and the others that are so useful to us. But a space traveller four and a half billion years ago would hardly have noticed the formation of yet another ordinary star, with nothing distinctive about it. Perhaps it would have stood out as a single star among the forest of double and triple stars. It also possessed a dusty disk. In a few million years, when our time traveller passed by again, the disk had vanished and in its place was a set of eight different planets and 60 assorted satellites surrounded by a cloud of comets. Whether this sight was unusual, or common and whether copies of any of the planets exist elsewhere, is the topic of this book.

Stars and planets: what is the difference?

We see that in our solar system the planets formed in a manner distinct from that of the Sun. A few years ago we were reasonably sure that this would apply throughout the universe. Now we are not so sure. Other planetary systems look different. Other disks of gas and dust may break up in different ways, depending on their size and how fast they are spinning.

Stars are simpler than planets, at least from the viewpoint of a planetary scientist. The development of stars can be studied during most of their evolutionary stages. Astronomers thus possess a considerable advantage compared with planetary scientists, and even with students of history: they can look backwards in time. As the universe is large enough they can find many samples both of stars and galaxies at more primitive stages in their evolution. The origin of stars about the size of the Sun is reasonably well understood. However, we are no longer so certain about their tiny relatives that overlap in mass with our larger planets.

Figure 7. Part of the Eagle Nebula (M 16) 7000 light years from Earth. The spectacular pillars are about one light year long and are produced by erosion of a molecular cloud by ultraviolet light from nearby young stars. The denser globules at the tips of the pillars are probably sites of the formation of new stars. (Hubble Space Telescope photo, courtesy P. Scowen, J. J. Hester and NASA.)

Stars form by condensation of gas that has separated from dense clouds of gas and dust (see Figure 7). As the gas contracts through the force of gravity to form a star, the temperatures and pressures rise to the point where the nuclear furnace ignites, and hydrogen begins to be converted to helium. This provides an enormous source of energy, which provides the familiar sunshine that we all enjoy. The heat produced balances the crushing force of gravity that is trying to collapse the star. Like a fire, the star can survive as long as it has fuel.

Before it settles down to a long steady middle age, the star undergoes some quite violent behaviour, like most other youthful creatures. These early excesses are labelled with the entertaining names of T Tauri or FU Orionis stages, named from examples of young violently acting stars. It is generally thought that the early Sun went through similar episodes.

The lifespan of stars is well understood. For example, the nearby star Beta Hydri, which is very similar to the Sun, is about nine billion years old, or twice the age of the Sun. This reassuring observation provides us with the comforting assurance of a long life for the Sun and thus for the solar system. But we also see the catastrophic end of stars. Towards the end of its life, which I talk about later, the Sun will again become violent, as though seized by a kind of dementia.

Stars have to be about 80 times the mass of Jupiter before the central pressures and temperatures are sufficient to cause the nuclear furnace to ignite. The Sun, undistinguished among a host of similar stars, is over 1000 times more massive than Jupiter and shines well, as we all observe.

Single and double stars

Single stars are not very common, and are perhaps even rare, compared with double star systems. These pairs form most of the stars that we see. Even triple stars are common and over three quarters of all stars live in double or triple associations. Although it is often supposed that the Sun and Jupiter represent a failed double star system, this idea neglects the fundamental difference between the processes responsible for making planets, as opposed to stars, in our solar system. As will become clear later, Jupiter is not a failed star that formed by condensation from a gas cloud. It is a true planet, which was built up bit by bit.

I need to look a little at the problem of forming double stars in order to separate this process from that forming planets. There are a number of classical explanations for the formation of double stars that remind us of the old explanations for the origin of the Moon. The simplest model involves capture of another star. Another breaks up a star into two bits by fission. Yet a third idea simply links two stars that formed close together, like a marriage between people from the same village. There are problems with all these models. The capture model does not explain why double stars are often close in mass to each other just as giants don't usually associate with pygmies. One would expect that captured companions might exhibit a great variation in size. Fission, in contrast, needs the parent star to rotate very rapidly so that it can fly apart into two pieces. It's difficult to get the star spinning rapidly enough. Finally, the binding together of stars that formed separately scarcely constitutes an explanation. In fact, double stars appear to form rather easily from the collapse of rotating gas clouds that evolve into shapes like dumbbells.

What makes a double star system form in the first place rather than a single star and a planetary system? The answer seems to be connected both with the mass and the speed of rotation of the fragment of the molecular cloud. Thus whether a single star or a double star forms is essentially random. It is merely a consequence of the size and amount of spin possessed by the initial cloud fragment. It is on such matters that the ultimate existence of the solar system, and of ourselves, turns. A larger faster spinning cloud would have produced not a central sun and planets, but a double star. In that event we would probably not be here to consider the problem. Although planets perhaps form around double stars (at least one apparent example is known), the complex orbits are not likely to produce a very congenial environment for intelligent life to develop, because of wild variations in temperature as the planet approaches to or recedes from its two suns.

Building stars

The formation of normal stars is one of the classical problems in astronomy. Although well enough understood, there still remain some unsolved problems dealing with the birth of stars. Stars form at very different rates in different types of galaxies, depending on the amount of gas available. Our galaxy, like most spiral galaxies, seems to have been forming stars at a rather regular rate for a long time. Many elliptical galaxies have used up all the available gas and star formation has ceased. Others, particularly where two galaxies have collided and a fresh supply of gas has been acquired, produce stars in a sort of frenzy. These are appropriately labelled 'star bursts'. However, the basic process of star formation seems to be the same. It is the rate that varies.

Stars form after fragments break away from the dark clouds of gas and dust that occur commonly in the spiral arms of our galaxy

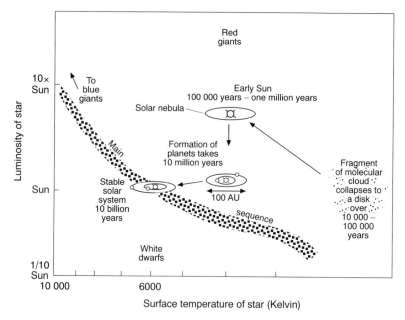

Figure 8. The formation of the Sun and planets, starting with the collapse of a fragment of a molecular cloud to a disk (the solar nebula). As the central star forms and heats up, it finally arrives on the 'main sequence' of stars, where the Sun and planets are stable for about 10 billion years. This sequence is shown superimposed on the famous Hertzsprung–Russell diagram of temperature against brightness of stars.

(Figure 8). These so called 'giant molecular clouds' are the most massive objects in the galaxy. They are composed of many smaller clumps, and are perhaps a hundred or so light years in diameter. They contain enough gas to form a million stars. The classic example of a giant molecular cloud is the Orion nebula. This forms the middle 'star' of the sword of Orion, the Hunter. (It is an historical accident that the term nebula is used both for these immense clouds and for the very much smaller disk of gas and dust, the solar nebula, from which our planetary system formed.) In these giant clouds, nature has managed to form many complex molecules, mostly of hydrogen, oxygen, nitrogen and carbon. Over 100 different organic compounds have been identified. There is enough ethyl alcohol, the common intoxicant, out there, not only to sink a ship, but to drown the Earth.

Many of the clouds contain small dense clumps. These are typically a tenth of a light year across, and are about the mass of the Sun. These clumps are the site of star formation. It takes about a hundred thousand years for the gas to collapse due to gravity to the point where the pressure and temperature are high enough to turn on the nuclear furnace.

Why do stars not get infinitely big? Such simple questions (why are most stars about the same size?), like other similar matters (why is the sky dark at night?), often conceal fundamental truths. There are good reasons why stars stop growing. They drive away the gas in the disk around them as they start to shine. What stops the gas falling inwards? The onset of strong stellar winds blowing out from the star is the answer. These winds reverse the flow of material into the star. The sizes of stars are thus limited and they do not grow to an infinitely large mass. The ignition of the nuclear furnace occurs when the mass reaches about one third of the mass of the Sun. Strong outflows of gas and jets of material then break out from the turbulent young star. This stops the infall of gas. The gas and dust now form a spiral disk around the new star. Eventually, a central star is left surrounded by a rotating disk from which some planets might form.

We see evidence of these early strong winds in the T Tauri and FU Orionis stars. These young stars are less than a million years old. They are important because they tell us what the early Sun was like and so cover a crucial period in the development of the solar system. Many are surrounded by dusty disks from which planets may form.

Curiously enough, it has been realised only recently that the Sun and other stars are composed almost entirely of gas in the form of hydrogen and helium. The high abundance of these gases in the Sun and in other stars was confirmed only around 1925. It is interesting that such a fundamental fact about the universe is such a recent discovery. The Sun contains only about two per cent of elements heavier than helium. In the astronomical world, all these, including such different elements as chlorine, nitrogen, oxygen and sulphur, are referred to as 'metals', to the exasperation of chemists. The story of the formation of the chemical elements, worked out through a combination of nuclear physics, astrophysics and astronomy in the 1950s, is one of the great triumphs of human understanding of the universe, which can only be mentioned here in passing.

The fate of the Sun

In about another five billion years, the Sun will eventually come to the end of its life. As the hydrogen in the core is used up, gravity will start to take over as the furnace shuts down. As the Sun collapses, the pressure inside builds up to the point where the furnace reignites, burning hydrogen in a shell outside the core. In the course of this, the Sun swells up to become a red giant, expanding out to the orbit of Venus within a few million years. However, it will have shed perhaps a quarter of its mass in the process. The Sun shrinks in volume again as the fire goes out. As the temperature in the core reaches 100 million degrees, a second cycle of nuclear fusion, involving helium, begins. The Sun balloons out in a second red giant stage. During this time, such useful elements as carbon and oxygen are produced in the fiery furnace. Further catastrophes and flashes follow, during which the Sun sheds most of its mass. The elements it has so usefully produced are scattered into space, providing material for new stars. As the furnace finally shuts down for want of fuel, gravity will exert its overwhelming force and reduce the Sun to a geriatric white dwarf, about the size of the Earth. This tiny creature will be so dense that one cubic centimetre will weigh many tons, a challenge beyond the imagination of our weight-lifters. The Sun is not massive enough to collapse into a black hole. That ultimate fate is reserved for much more massive stars. In the end, our brilliant Sun will become an invisible black dwarf.

Red dwarfs, brown dwarfs and possible tiny relatives

Red dwarfs are stars that are smaller than the Sun, ranging down to about one tenth of the mass of the Sun. Those that are a bit smaller than the Sun are very common. Perhaps 80 per cent of all nearby stars are red dwarfs. However, smaller ones seem to be rarer. Their numbers appear to fall off rapidly for stars with less than 20 per cent of the mass of the Sun. However, like much else in science, this decline may be more apparent than real, due more to the difficulties of observing these small bodies than their absence.

Between the smallest red dwarfs and large planets such as Jupiter lies the realm of the brown dwarfs. They are of interest here because some of the 'new planets' may be related to them. Brown dwarfs are stars, but are very cool by stellar standards because they are too tiny to reach temperatures and pressures high enough to 'burn' hydrogen to helium. Like kiwis in my native New Zealand, brown dwarfs remain very elusive. As with other rare species, a lot of effort has been spent in looking for these creatures. Good hunting grounds for them are to be found in the young star clusters. One is the Hyades cluster, a beautiful group in the constellation of Taurus, the Bull, which lies between the Pleiades or Seven Sisters, and the magnificent constellation of Orion, the Hunter. This cluster consists of several hundred stars lying about 130 light years from the solar system. The stars in the Hyades formed about 600 million years ago, just as life on Earth was undergoing an explosive diversity of life forms and the first hard-shelled animals were making their appearance. These we see as trilobites and other species preserved as fossils in strata of Cambrian age, as well as the marvellously preserved soft bodied forms of many extinct animals in the Burgess Shale, which was then mud on the ocean floor, but which would eventually become part of the Rocky Mountains of British Columbia. The Burgess Shale will appear again in the story.

The splendid group of the Pleiades, about 400 light years away, is another good prospect for these elusive fellows. The stars in the Pleiades formed about 100 million years ago, when the dinosaurs were enjoying their heyday in the warm Cretaceous sunshine. It was also around this time when a large asteroid or comet hit the Moon, forming the great crater called after Tycho. This crater is noteworthy for the spectacular set of rays of dust flung out by the explosion, which go right across the face of the Moon. They are easily visible in a small telescope or binoculars.

However, only a couple of potential brown dwarf candidates, with the pedestrian designations of Teide 1 and PPL 15, have been found in the Pleiades, while searches in the Hyades cluster failed to locate more than one or two more doubtful examples. This was much less than the expected swarm, or whatever the proper collective noun is for dwarfs. A wider survey by an infra-red satellite, which would have been expected to find such cool stars, failed to locate any. One firm discovery was made much closer to home. Gliese 229B, about 50 times the mass of Jupiter, is a faintly glowing companion of a red dwarf star (Gliese 229A) that is only 19 light years away. At long last, we have a candidate that everyone is agreed upon.

There is a lot of debate about the smallest mass of gas that can collapse to form a tiny star-like body. We used to think that the smallest clump that can form from an interstellar cloud is about ten times the mass of Jupiter, but perhaps smaller clumps are possible. One property may be useful to tell the difference between large planets and brown dwarf stars. Planets are more liable to be in near-circular orbits, while the dwarfs are likely to be on highly eccentric orbits, like most double stars. On this basis, most of the discoveries of the larger new 'planets' around stars may be brown dwarfs. Nature thus seems to have produced a wide variety of tiny fellows. All this signals the uniqueness of our own solar system. I return to the discussion of other planetary systems towards the end of this story, after I have traversed the marvellous complexity of our own.

Finally, getting back to cosmology, the brown dwarfs are rare enough that they are unlikely to make a significant contribution to the mass of the galaxy. If these creatures were numerous, then they could contribute significantly to the famous problem of the missing mass in the universe, but they are clearly not the solution.

Disks around stars

Perhaps half of young stars (with ages less than 3 million years) that have been surveyed are surrounded by dusty disks similar in size to the solar system. By the time that stars are more than a few million years old, the gas has gone. This places strict limits on the formation of gas giants such as Jupiter. Clearly, such planets must form quickly before the gas is lost.

The best example is the disk around the young star HL Tauri, where we may be looking at the beginning of a planetary system. The disk is rotating. It is about one tenth of the mass of the Sun and has a diameter of about 2000 AU. Other disks were discovered around Alpha Lyrae (Vega), Beta Pictoris, Epsilon Eridani, and Alpha Pisces (Formalhaut). These are middle-aged stars. The typical size of their disks is 200 AU. The Beta Pictoris disk has about an Earth mass of gas and dust and could be similar to the cloud of comets that surrounds our own solar system. It appears to have gaps and warps, possibly

Figure 9. The dusty disk which extends out to about 400 AU around the star Beta Pictoris. The disk has a thickness of about 50 AU at 300 AU from the star. Light from the central star is masked (courtesy B. A. Smith, University of Arizona).

indicating the presence of planets. This discovery that dusty disks are present around distant stars (see Figure 9) leads us to consider whether there was one around our early Sun, a topic to which I now turn.

The disk around the Sun

Laplace and his solar nebula

In the eighteenth century, scientists were unaware of the backward revolution of Venus, of the existence of minor satellites with exotic orbits, of the strange orbit of Pluto and of other irregularities. The solar system that they saw appeared as well-ordered as a clock. The planets and satellites then known lay close to the plane in which the Earth rotates around the Sun. They also rotated in the same sense, both around the Sun and about their axes of rotation. This fortunate lack of too much information enabled the French astronomer and mathematician whom we have met before, Pierre Simon, marquis de Laplace, to propose in 1796 that the solar system originated from a rotating disk of dust and gas (as I noted earlier, he ignored the strange orbits of the satellites of Uranus). He called this disk the *solar nebula*. In his model, the planets condensed successively from rings as the nebula contracted. This concept survived in its original form until late in the nineteenth century. The view that the Sun and the planets formed from a rotating disk of gas and dust, the solar nebula, now provides such an obvious explanation that it has become an ingrained truth. Laplace would no doubt have been pleased that his concept has survived.

The ancient Greeks needed five components to build the heavenly bodies and the Earth. We have made some progress. The great mixture of ingredients that were in the original disk and from which the Sun and planets were built can be reduced to three: gas, ice and rock (see Figure 10). The gas was mostly hydrogen and helium, the two

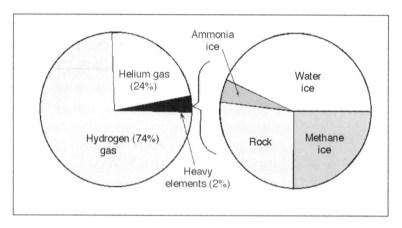

Figure 10. Gas, ices and rock are the three main components of the primitive solar nebula. These pie diagrams show their relative abundances.

elements that form most of the universe. The gas made up 98 per cent of the mass of the initial solar nebula. The remaining two per cent consisted of various ices and rock composed of the heavier elements. The ices were mostly water ice, ammonia, nitrogen, carbon monoxide and methane (the temperatures were low enough so that these compounds were present as ice). Rock is most easily imagined as something like the stony meteorites, a mixture of silicate minerals, sulphides and metal. The metal is mostly iron alloyed with some nickel and cobalt.

The planets themselves likewise fall into three classes. The gas giants, Jupiter and Saturn, are mostly gas. The ice giants, Uranus and Neptune, are mixtures of ice and rock with only a little gas. Our familiar terrestrial planets are mostly rock with cores of metal. How did we finish up with our Earth and all the other marvellous complexity of our solar system from such unpromising material? How all this came to pass is a major theme of this book.

In the beginning

The disk from which the Sun and planets formed has now vanished, reminding one of the Cheshire Cat in *Alice in Wonderland* that disappeared, leaving only its smile behind. Our attempts to reconstruct what the nebula was like originally suffer from the same problems that historians have. One has to avoid both folklore, overinterpretation and wishful thinking. Clearly, most of the original nebula finished up in the Sun. What concerns us here is how much was left over to form the planets. This disk is usually what people are talking about when they use the term solar nebula.

The solar nebula began when a fragment broke off one of the molecular clouds that I talked about earlier. What causes the clouds to break up? What determines the size, rotation and spin of the fragments? Why do some bits finish up as double stars? What distinguished our own nebula from larger or smaller ones? These questions remain fertile fields for investigation. Shock waves coming from supernovae have been a popular suggestion for a way to break up dense gas clouds. However, the clouds perhaps simply collapse under their own gravity once they become cool enough.

How big a disk?

Historically, there have been two competing models for the original size of the solar nebula. One model for the primitive nebula has it containing about double the mass of the Sun. In this model, half the mass vanished into the Sun and the planets were made by breaking up the disk into a number of fragments. These condensed into giant puffballs, called giant gaseous protoplanets. This model has the great advantage of forming the giant planets more or less instantly during the brief lifetime of the nebula. This is the most agreeable feature of the model. However, one is left with a major disposal problem, which dwarfs any of our terrestrial experience in such matters. Over 99 per cent of the material has to be thrown away. There are other problems that I talk about later.

The alternative model supposes that the original disk from which the planets formed may have been quite small. The planets and everything else in the solar system now amount to only about one thousandth of the mass of the Sun. The lower limit to the original size of the disk is obviously given by the present masses of the planets. One must add in some extra gas to compensate for that lost from the region of the inner rocky planets and the hole at the asteroid belt. In such models, the thin nebula behaves very differently from a massive one. It did not break up into planetary-size bits of its own accord. Instead, the dusty grains of rock and ice separated from the gas and settled toward the centre plane of the disk. Boulder-size bodies grew by sticking the particles together and larger chunks formed by collisions. Eventually, quite large bodies, the size of mountains, arose. These looked like our asteroids, and are called planetesimals, a term that will occur from time to time throughout the book.

The justification for using this name is that the current model for building planets, at least in our solar system, is known as the planetesimal hypothesis. It has a respectable history, dating back to the beginning of this century. This planetesimal model effectively built the planets up brick by brick. Thus, it is in great contrast to the previous idea that broke up up the disk into giant puffballs of gas. So one can make the planets out of a much smaller disk, perhaps only about 10 or 20 times the mass of Jupiter. Even so, quite a lot of material gets tossed away.

A short life

For how long did the solar nebula last? Clearly, one has to form the Sun and gas-rich planets while the gas was around. The time from the initial separation of the disk of gas and dust from the molecular cloud out in the galaxy to the point where the Sun is large enough to ignite the nuclear furnace is somewhere between a hundred thousand and one million years. Once the Sun begins to shine, the gas left over in the nebula is soon driven away. Although we see that dusty disks (of ice and rock) persist for periods of a few million years around violently behaving young stars such as T Tauri, the gas itself may have been lost over a much shorter lifetime, perhaps as brief as one million years. The time between the formation of the solar nebula as a rotating disk of dust and gas and the disappearance of the gas is very short. Thus, the gas giants, Jupiter and Saturn, have to form rather quickly, before the gas has been driven away. As will become apparent, some fine timing is needed to form Jupiter at all.

Was the disk hot or cold?

Some detective work is needed to reconstruct events that occurred over four and a half billion years ago. We can of course get some ideas from the present condition of the solar system. The inner planets are rocky, without much gas or ice. Gas is now present far away from the Sun at Jupiter. The satellites out there have lots of water ice, except for the special case of Io, which is cooked by being so close to Jupiter. Closer in, we have samples from the region between Jupiter and Mars. These are the meteorites that come from the asteroid belt. They date back to the beginning of the solar system and have a special significance since they tell us about temperatures at that remote epoch. Out at about three AU, it was hot enough to melt ice. The meteorites that were a little closer to the Sun have been in a hotter zone. They have lost not only water, but varying amounts of elements like lead, potassium and other easily vapourised elements.

So clearly the nebula was getting hotter nearer the Sun. Just how hot is a question. Primitive meteorites are complex mixtures of minerals formed at low and high temperatures. About half of our stony meteorite samples consist of tiny glassy spheres, typically about a millimetre in size. These are the famous 'chondrules' discovered over a century ago by Henry C. Sorby (1826–1908). He was a gentleman scientist in the Victorian tradition, who invented the powerful technique of examining thin transparent slices of rock under a microscope. When he turned his attention to meteorites, he recognised that the chondrules had been 'molten drops in a fiery rain' that had cooled to glass. We have not made too much progress since then. Human ingenuity has suggested all possible processes without much agreement.

The chondrule factory must have been efficient, for at least half of the material in meteorites seems to have passed through it. It is clear that chondrules had been dust balls that were flash melted. As they cooled very quickly, the part of the disk where they formed could not have been hot all over. They had to melt and cool very quickly, so the nebula was not uniformly hot.

Clearly the nebula was not a cold, inert disk, but a dynamic and turbulent system with much energy to dissipate. A large spiral storm system on Earth, called a cyclone, hurricane or typhoon, according to taste and local folklore, is a good parallel. In these storm systems, there are a lot of thunderstorms and local turbulence within the overall large spiral structure. They are probably quite a good model for conditions in the early disk that surrounded the Sun.

Such a model also explains why the Sun is spinning very slowly, taking about a month to turn around, and why the planets are spinning rapidly. This is an ancient paradox. It's very difficult to accomplish this with a symmetrical disk. Most naturally formed disks, like hurricanes or spiral nebulae, however, are not nicely symmetrical. If the solar nebula resembled these uneven disks, then, as mass fell into the Sun, the spinning disk could spread out, accounting for the fast spin of the planets

Before the solar system

Ben Jonson (1573–1637), the Elizabethan author, wrote with great insight in his play *The Alchemist* that it was 'absurd to think that Nature in the earth bred gold, perfect i' the instant – something went before. There must be remote matter'.[8] We are now well informed about remote matter. As I mentioned earlier, one of the more impressive achievements of science in this century has been the explanation of the origin of the chemical elements. Hydrogen, helium, deuterium and a trace of lithium were formed in the Big Bang. The heavier elements, which the astronomers persist in lumping together as 'metals', were made in an immense number of stars for over ten billion years, and were scattered out into interstellar space.

The two main sources of this scattering were, and are, red giants, which shed a lot of mass out into space, and the great stellar explosions that we see as supernovae. During such stupendous events, some mineral grains were formed on the shock fronts as the explosion radiated outwards. Little is known of the conditions to which the gases and grains were subjected on their long journey from supernovae to the solar system.

What is remarkable is that some interstellar grains that are found in meteorites have survived all the hazards. They are identified by variations in the abundances of isotopes that are bizarre by the standards of the solar system. By this means, diamond and silicon carbide that formed long before the solar system have been identified in meteorites. The diamonds are tiny, typically containing only 25 atoms or so. As Ed Anders (b. 1928), who discovered them a few years ago while working at the University of Chicago, remarked, they are the right size to form the stones if bacteria wore engagement rings. These diamonds formed on the outskirts of supernova explosions, although the exact process remains unclear. The silicon carbide was made in red giant stars as they swelled up and, like celestial strippers, shed their outer envelopes. These grains of diamond and silicon carbide survived the rigours of interstellar travel from the supernova or red giant to the solar system on a scale that would astound the crew of *Star Trek*. Once here, they were caught up in the turbulent events in the early nebula. It is remarkable that they remain to tell the tale of their distant origin, before the solar system began.

What was the disk made of?

It may seem surprising that we are well informed about the composition of the primitive solar nebula, which vanished so long ago into the Sun and planets. Although it was chiefly composed of gas, the main interest here revolves around the two per cent of rock and ice, partly

because we live on some of it. The Earth, however, has had a complicated geological history. The rocks that we now find on the surface
are the end product of four billion years of recycling and now contain only a faint record of their primitive beginning. Fortunately,
some samples from these remote times have been delivered to us.
Meteorites frequently fall on the Earth, mostly having been kicked
out of the asteroid belt, of which more later on. Among this debris
from space, we are fortunate enough to have found some meteorites
that are unchanged from the earliest times. Some have compositions
that closely match the rocky component of the original solar nebula.

How do we know this interesting fact? Because these meteorites,
which come from beyond 3 AU, have the same ratios of elements such
as sodium, iron, magnesium and uranium that we see in the spectra
that are recorded from the Sun and tell us of its composition. With
trivial exceptions, the match is good for all elements that made up the
rocky component of the original nebula.

Probably a small digression on some differences between the
behaviour of the chemical elements in the nebula and in our high
school chemistry laboratories is in order here. Everyone is familiar
with chemical reactions on the surface of the Earth. Mainly it's a matter of forming compounds such as sodium chloride that every schoolchild knows about. The grouping of elements in that great triumph
of nineteenth century science, the Periodic Table of the Chemical
Elements, explains why, for example, sodium atoms join with those
of chlorine to form table salt. It's a matter of how the outer shells of
electrons of the atoms are arranged. These properties are less important at temperatures hundreds of degrees hotter than the laboratory
Bunsen burner, and pressures lower than the best vacuum that we can
obtain.

Out in the nebula, other properties take over. The most important
is whether the elements have high or low melting and boiling points.
Those that melt and boil at high temperatures are called refractory.
For the chemist studying the cosmos, elements that boil above 1200 K
or so are refractory. Examples are titanium, calcium and uranium. At
the lower end of the scale lies the region of the volatile elements,
which includes such common examples as lead, potassium, sulphur,
copper and zinc. Although they are all nicely stable in our chemical
laboratory, when temperatures got high enough in the primitive

nebula, these elements evaporated, leaving only the refractory elements in solid form.

When solid compounds begin to form, two other factors become important. The first is the relative size of the atoms, which determines whether they can fit into crystals or are pushed aside, like square pegs trying to fit into round holes. The other chemical elements that are present mostly in trace amounts have to scramble around trying to find the crystals with the appropriate cavities that will take them in. The other factor concerns which particular minerals are made. As the dust grains condense, three main species of minerals form. These are silicates, sulphides, and iron metal. The minerals are mostly made of the common elements such as iron, magnesium, aluminum, silicon and sulphur. Here endeth the first lesson on chemistry out in space.

One might have thought that our Earth would be a prime example of the original rocky component of the nebula. It turns out, however, that the inner rocky planets, Earth, Venus and Mars have lost some volatile elements, such as sodium, potassium and lead. How this comes about will appear later.

Was the disk uniform?

One might also confidently have expected the primitive solar nebula, a fragment from a molecular cloud, to be nicely uniform in composition and indeed this used to be the common perception. This topic has had an interesting history. It reflects a common approach to many scientific problems. What we cannot see or measure, we imagine to be simple and homogeneous. The approach is related to the human tendency to underrate what we do not understand. This intellectual safety valve allows us to live with the unknown, and is no doubt responsible for the popularity of religions, astrology, mysticism and the other engaging fantasies that the human mind has dreamed up in the absence of facts. These provide nice stories, like the flat Earth, that are obvious to children. Reality is a bit more complex.

Surprisingly, the nebula was not very uniform. The most significant observation was that in samples from different meteorites, oxygen, which is one of the most common elements, has variations in the ratios of its three isotopes. If all the material in the primordial nebula had been heated and vapourised, then these isotope ratios would be the same. The grains of diamond and silicon carbide would not have survived. So there must have been much local variation.

The sweepout of gas

Although gas made up 98 per cent of the original solar nebula, we are standing on a rocky planet that has only a trace left. Mercury and Mars are in worse shape, and even cloud-shrouded Venus has only a little more gas than the Earth. When did this drastic decline happen? There is strong evidence that young stars lose their disks of gas within a few million years following the birth of the star. This presents us with an interesting problem. The gas giants, which are far from the Sun, must have formed while the gas was still around. In great contrast, the Earth and the other inner planets were put together after the gas was gone.

So we begin to get some clues about when the planets formed, a topic that I address shortly. The big gas planets had to form early, within a few million years of the formation of the Sun. The inner rocky planets were assembled later from the left-over bits and pieces after the gas was gone.

The inner parts of the nebula were thus cleared of the gas and depleted in volatile elements very early, probably within about one million years of the formation of the Sun. The cause of this clearing of the inner regions of the nebula was twofold. Early on, gas was swept into the Sun. Then, after the nuclear furnace ignited, strong winds blew outward from the Sun and swept away any remaining gas. A few bodies, ranging in size from boulders to small mountains, survived even the very strong stellar winds. This was a fortunate circumstance, because we are standing on a planet that formed from this collection of rubble.

The finer material, dust, smoke and gas was swept away. Along with the gas went varying amounts of the volatile elements that had not condensed or found refuge in solid minerals. We even have, courtesy of our useful meteorite samples, a date for this event. This is arguably their greatest contribution to our knowledge of the early solar system. The meteorites contain elements with radioactive isotopes that provide clocks that can be read. Thus, some isotopes of lead, a volatile element to the planetary chemists, are produced from the radioactive decay of uranium and thorium. These are both refractory elements with high boiling points of a few thousand degrees. Another example is the volatile element rubidium, a sister element that is inseparable from potassium, which has one isotope which undergoes radioactive decay to the refractory element strontium. Much of the volatile lead and rubidium were swept away with the gas, leaving the refractory elements uranium, thorium and strontium. Reading these radioactive clocks tells us that there was a great separation of volatile from refractory elements in the inner solar system almost back at the beginning of the solar system. The agreed date is 4566 million years ago. We know this time to within two million years or so. This astonishing accuracy is a tribute to painstaking and laborious work. Two famous laboratories engaged in these studies are at the University of Paris and at the California Institute of Technology. The latter is widely known as the Lunatic Asylum. Like such institutions, it is characterised by security and the presence of people in white coats.

Building planets

The collapse of clockwork solar systems

The orderly system that was represented by the clockwork models influenced thinking about origins of the system until very recently. Theories of a simple hot nebula, cooling down and condensing into the planets in a regular fashion, were popular until only a few years ago. The concept of a heavenly clockwork, illustrated by the well-known picture of a man peering past a veil of stars to discover a sort of grandfather clock mechanism beyond, typifies this viewpoint. The clockmaker presumably lurks behind the clockwork.

Even so distinguished and skeptical a scientist as Lord Kelvin (1824–1907) was led to make the following comment: 'There may in reality be nothing more of mystery or of difficulty in the automatic progress of the solar system from cold matter diffused through space, to its present manifest order and beauty, lighted and warmed by its brilliant sun, than there is to the winding up of a clock, and letting it

go till it stops. A watch spring is much farther beyond our understanding than is a gaseous nebula'.[9]

The search for meaning and for regularities and order in the
system probably accounts for the popularity of Bode's Rule. The
regular spacing of the planets looks like part of a grand design. But
the 'rule', as I discussed earlier, is a minor result of tidal forces
between planets. It is not a property of fundamental significance nor
is it part of a grand blueprint for constructing planetary systems.

Such views of a tidy and well organised solar system have not survived. No two planets or satellites are alike. As in so many other fields,
ranging from astronomy to genetics, we are reluctantly compelled to
realise that we inhabit a system in which chance events play a major
role. Such ideas of the randomness of nature are unpopular. They run
contrary to the egocentric philosophies in which *Homo sapiens* occupies a central role in the universe and in which all is apparently
designed for our comfort and well-being. The reality is different. The
solar system is just another physical system in which random events
can occur. This recognition of the importance of chance events has
been one of the more profound changes in our perception of the
world since the construction of a clockwork solar system could be
ascribed to a divine watchmaker.

The problem

The problem of building planets is fundamental to the entire question of the origin of the solar system. Historically, this latter question
has frequently been considered to have been answered, but the wide
variety of explanations and solutions that have been offered, from the
creation myths of primitive societies, to the more recent, but numerous scientific attempts, has generally collapsed when faced with new
information about the system.

There are two principal difficulties. The first dilemma is that the
planetary scientist, like the historian, has only one example, the present scene, together with whatever relics have survived from previous
epochs, to tell the tale of former events. One must of course be skeptical about relics. There is a long history of fraudulent relics of which
two recent examples are Piltdown Man and The Shroud of Turin.

Statistical treatment has been successful in dealing with the formation of stars, of which there is a great abundance. However, statistics are of little use in understanding a single system, since improbable events can always happen once. This is usually illustrated by the tale of the only elephant in the Leningrad (now St Petersburg) Zoo. During the siege of the city by the German Army in the Second World War, they deployed a giant gun, capable of firing over 20 kilometres into the centre of the city. The first shell hit the zoo and killed the elephant. Other versions blame the first bomb to fall on the city. However, the fate of this unfortunate beast may be a fable, for the story is not mentioned in the definitive accounts of the 900 day siege.[10]

A major problem arises because of the unknown initial state of the solar system. We see only the final product, and have to infer both the way by which it came about, and the starting point. Like the path of terrestrial biological evolution, both trails are non-unique. How could one, when contemplating an elephant, deduce the existence of bacteria, or DNA for that matter. The wide variety of explanations that have been proposed to account for the solar system reflect these uncertainties.

A central philosophical point is whether the planets condense from fragments of the nebula or are built up by accumulation of smaller particles. There are only two basic ways to make things. The first is to start with something big and break it into pieces. The second is to build a larger structure from smaller bits. An example of the first process is carving a statue out of a piece of marble; of the second, building a house from bricks. Such thoughts have resulted in two contrasting models for building planets. The first process breaks up the gaseous solar nebula, from which planets condense in the same way as stars. The second process builds planets up out of smaller bits and pieces.

It is perhaps worth commenting on two obvious truisms. First, the solar system is very isolated. The nearest stars are so far away as to have only a trivial influence during the construction of the system. Second, the whole system effectively lies in one plane, with most of the bodies orbiting the Sun and rotating in the same direction. This was what had impressed Laplace. The flatness of our system does seem to be a primitive feature, inherited from the rotating disk-form of the solar nebula. Both these factors constitute evidence for a common origin of the Sun and the planets, as Laplace noted. Both arrangements are unlikely to result from a random accumulation of objects. Otherwise, a case could be made for assembling this heterogeneous collection of different planets and satellites from some kind of cosmic junkyard.

This idea of a common origin for the Sun and planets has been part of the general agreement about the solar system for the past 200 years, ever since it formed the basis for the Laplace hypothesis. However, the junkyard analogue would seem to fit the new 'planetary' systems that we have found, for no two are alike. They show many differences among themselves and none look remotely like our system. The new information from these very limited examples, which I talk about later, seems to confirm the uniqueness of our own system.

Planetary formation, at least in our system, is a very wasteful and inefficient process. Much of the nebular material either finished up in the Sun or was thrown away into outer space. Even in the planetesimal model, advocated here, the planets have finished up with less than ten per cent of the material originally available. Earlier models, such as those making giant gaseous puffballs from a massive disk, had to throw away 99 per cent of the material originally available. What would one think of a factory that threw away most of the raw material? So little material finishes up in the planets that an accountant might write it off.

Three types of planets

The most striking difference in the solar system is the distinction between the giant planets and the small rocky inner planets. There is really a threefold division, because Uranus and Neptune are ice giants rather than gas giants like Jupiter and Saturn. The planets themselves thus mirror the three main constituents of the solar nebula, gas, ice and rock. In addition to these differences in composition between the three groups of planets, there is also the striking difference in mass between them.

There is little reason to suppose that this threefold division would be repeated in another planetary system, as it seems to depend on a number of chance events, starting with the size of the fragment of the molecular cloud that broke away to form the Sun and planets. As I discuss towards the end of this account, a larger nebula might produce three rather than two gas giants. In that case, models predict that one is thrown right out of the system, another is sent far away on a wildly eccentric orbit, while the third finishes up in close orbit around the star.

Giant puffballs

We have met these before. One way of making giant planets or brown dwarfs is directly, by breaking up the solar nebula. Because such a process would occur rapidly, it has the advantage of making gas-rich planets or dwarf stars while the gas is still around. This advantage, however, is offset by many other problems.

The models that build such gas-rich planets have some interesting and predictable consequences. The principal attraction is that it quickly breaks up the nebula into Jupiter-sized bodies, thus accounting for the giant planets. But if the nebula breaks up into fragments, all planets should exhibit similar compositions or exhibit some regular and systematic change with distance. The problem is that the giant planets are all different. The proportion of gas varies among all four giant planets. Jupiter does not have the composition of the Sun, as it should if it was simply a bit of the original nebula.

Another problem, equally fatal to the concept that Jupiter and Saturn are unaltered bits of the nebula, is that they have central cores of rock and ice. The existence of these cores is deduced from gravity measurements by spacecraft. There is some uncertainty in their size, with values typically ranging from six to 20 Earth masses. Here, I adopt ten as a reasonable average of the various estimates. How do these cores form? In the Earth, we have an iron core. This is because iron is so much denser than rock that it fell into the centre at the beginning when the Earth was molten. However, the pressure and temperature at the centre of Jupiter is around 70 million atmospheres and 20 000 K. Saturn is smaller, with a central pressure of about 40 million atmospheres and a temperature a little over 10 000 K. These conditions are so extreme that if the planet had condensed from a chunk of the nebula, everything would have remained in a homogeneous kind of soup and denser material could not separate to fall in to make a distinct core.

The only way to get a discrete core in such a large planet is have it there to begin with, and build the rest of the planet around it. In the planetesimal hypothesis, the core is built up from small bodies of rock and ice until it is big enough to collect an envelope of gas around it. I will discuss how this is done a little later.

For these reasons, the model of making planets bit by bit has replaced the concept of building the planets by condensing them straight from the nebula.

Did our planets accumulate from dust?

One might think that formation of planets from a dusty disk would occur simply by collecting the dust from the nebula around rocky centres. These might arise naturally through local accidents, or perhaps form from rings related to the Bode's rule spacing. This simple picture, which was a feature of some older models, does not fit the evidence. This tells us that planets accumulate mostly from large objects, not dust. A whole variety of bodies of differing sizes populated the early solar system, before being swept up into the present planets.

The battered faces of planets and satellites still display the impact of many large objects. The evidence is clear enough, visible on the Moon through a small telescope or binoculars. Basins the size of France or Texas, surrounded by rings of mountains, are common (see Figure 11). They are caused by collisions with bodies of the order of 100 kilometres in diameter.

The planets are mostly tilted, with their axes of rotation inclined at various angles to the common plane of the solar system. They mostly spin at different rates. This seems like a strange arrangement. If the planets had accumulated from dust, or from small objects only, they should all be in an upright position, since there would be nothing to disturb this tidy arrangement. Instead, they are tilted about as though they had been targets in a cosmic fairground.

An impressive collision is needed to tip the Earth through 24 degrees. The current model to explain the origin of the Moon calls for an object larger than Mars to collide with the Earth. This event may thus have produced the tilt of the Earth and also the 24 hour day. It takes an impact with a body the size of the Earth to knock Uranus

Figure 11. The site of a major collision on the Moon. Mare Orientale, which is about the size of France, was the site of a massive impact on the Moon 3800 million years ago, when a planetesimal about 50 kilometres in diameter slammed into the Moon and produced these concentric rings of mountains, several kilometres high, in a few minutes. Mare Orientale is the classical example of a multiring basin. The diameter of the outer mountain ring, Montes Cordillera, is 900 kilometres. The small dark area of mare basalt to the northeast is Grimaldi, visible from the Earth through binoculars. The western edge of the basalt plains of Oceanus Procellarum is on the northeastern horizon (NASA *Orbiter* IV 187 M).

over on its side. Venus is rotating slowly backwards. Why is this planet
so different? Perhaps it was due to a head-on collision with a Mars-
sized object that stopped it in its tracks and spun it backwards. But
perhaps Venus accumulated from small bodies, and escaped being hit
at all by a big one. Perhaps the upright stance and slow spin of Venus
is the natural result of it accumulating from a lot of small bodies.

Finally, accumulation from dust might be expected to produce
planets that are rather uniform in composition. Alternatively, some
regular differences in planetary compositions might be observed with
distance from the Sun. Neither of these signatures is apparent.

Thus the evidence from the present solar system tells us of the
existence of large bodies during the formation of the planets. The
planets assembled themselves from a collection of large bodies of all
sizes up to Mars and Mercury, not from dust.

The first solid bodies

Models that form planets by breaking up the gaseous nebula or those
that accumulate them from dust don't seem to work, at least in our
neighbourhood. We can now turn to models that build planets bit by
bit, like putting a house together from a pile of bricks. This is not without its problems. How do you start from a dusty disk with grains, typically a tenth of a micron in diameter, and finish up with a planet you
can walk about on that is so big that it seems flat to primitive societies?

What stuck the grains together? Ways to do so are not obvious.
Most models suggest that the particles adhere as the dust settles down
to the middle plane of the rotating disk. The problem is that as the
density of particles increases near the central plane, the disk becomes
turbulent. This turbulence prevents the particles from sticking
together, so a blind alley is reached. Despite these little-understood
problems, it is clear that solid bodies perhaps a few metres in size
must have formed within a million years of the formation of the Sun.
Otherwise, all material, gas and fine dust alike, would have been
swept either into the Sun, or out of the inner solar system. Having
nothing to stand on, we would not be here to discuss the problem.
Probably once the particles get big enough, perhaps a few metres in
size, they are no longer affected by the turbulence of the gas, and
clump together into bodies a few kilometres across.

How long does this process take? Estimates of a few tens of thousands to hundreds of thousands of years are usually quoted for this accumulation. Evidence for these very early events can be found in our informative meteorites. These have appeared from time to time in this account and now it is time for a short discussion about them

The most ancient samples

In 1492, the same year that Columbus discovered America, a meteorite fell near the village of Ensisheim in Alsace. This region is now a province of France, although then it formed part of the Holy Roman Empire. The fall was spectacular. The explosions as the body broke up in the atmosphere were heard for hundreds of kilometres as the body travelled north-westwards across Switzerland and exploded over the Rhine. The Emperor took this event as a favourable omen from Heaven and made a successful attack on the French. He ordered that the stone be preserved, and it is still there, in the Town Hall. The villagers put a chain around it, to prevent it flying back up into the sky. They also attached to it the following notice: 'Many know much about this stone, everyone knows something, but no one knows quite enough'. That is still fair comment about our understanding of meteorites.[11]

A decisive advantage is that we can analyse, and most importantly, determine the ages of meteorites. This is a non-trivial task that can only be accomplished in terrestrial laboratories. Some of the components of meteorites such as diamond and silicon carbide retain a memory of the time before the solar system existed. Other primitive meteorites match the composition of the Sun for many elements. Others tell us about the existence of small asteroids that melted and produced volcanic lavas within a few million years of the beginning of the solar system. Perhaps the most dramatic information is provided by the iron meteorites that formed in small parent bodies, less than a few hundred kilometres in size. The spectacular masses of iron that we see in museums, and that are everyone's idea of a meteorite, come from over 60 parent bodies. These were miniature planets, which melted a few million years after the beginning of the solar system. The iron sank down to make a core, surrounded by a rocky mantle. Some of our most spectacular meteorites are mixtures of

Figure 12. A meteorite found at Dora, New Mexico. It arrived on Earth after a small asteroid was broken up in a collision. Crystals of olivine from the rocky mantle are embedded in iron from the metallic core of the asteroid (courtesy Adrian Brearley, Institute of Meteoritics, University of New Mexico).

metallic iron and the green mineral olivine (see Figure 12). These came from the boundary between the core and mantle as the asteroids were broken up later in collisions that sent the bits and pieces tumbling earthwards. Nevertheless, meteorites need to be kept in perspective. They are rocks broken off asteroids and so come from a small, unique part of the solar system.

The crucial factor is that the record in the meteorites provides our only source of information about the earliest stages of the formation of the solar system. Thus they enable us to mark the the beginning of the solar system. It is always convenient to have a well-established base line to which history can be tied. The boundary between BC and AD forms one such marker for human history. However, if we were looking for a starting date for Western Civilisation, the date for the foundation of Rome, 753 BC, would have been a more logical choice.

What should serve for the solar system? Should we set zero time as the time of separation of the molecular cloud, the formation of the disk around the Sun, or the formation of the Sun itself? None of these events can be measured precisely. The most convenient marker for the beginning of the solar system is the age of the oldest solid bodies found in meteorites. These can be dated quite accurately. The oldest most reliably dated objects are the inclusions of refractory minerals. These are tiny grains of high temperature minerals that have been through repeated episodes of condensation and evaporation in the turbulent disk around the Sun. These grains formed 4566 million years ago, a date accurate, as I remarked earlier, to within a couple of million years. Although this was immensely long ago on human time scales, the universe had already been around for three times as long.

The building blocks

I promised earlier that the planetesimal hypothesis would reappear. According to the current version, the dust in the rotating disk around the Sun began to clump together at a very early stage. All sorts of bodies formed, beginning with grains that stuck together by obscure processes to form metre-sized boulders. In turn these grew to objects of kilometre size, finally reaching dimensions of hundreds or thousands of kilometres before they were swept up into the terrestrial planets. These early objects are the planetesimals I talked about before. Phobos, the larger satellite of Mars, is an example of such a body (Figure 13). Current thinking regards them as the building blocks of the terrestrial planets and of the icy and rocky cores of the giant planets.

The term *planetesimal* is due to T. C. Chamberlin (1843–1928) and F. R. Moulton (1872–1952), working at the University of Chicago in 1905. In their model, it referred to the small bodies that condensed from a filament pulled out of the Sun by a passing star. The model did not survive. Only the name has been kept as a useful term to describe the small bodies from which the inner planets were put together.

The basic problem with the model of Chamberlin and Moulton was that such a filament would have vanished like smoke into space rather than condensing into the planetesimals. There was yet another problem that became clear later. The planets have too much lithium, beryllium and boron, chemical elements that are rare in the Sun, to have come from there. These elements, although rare, are about 100 times more abundant in the planets and meteorites than in the Sun.

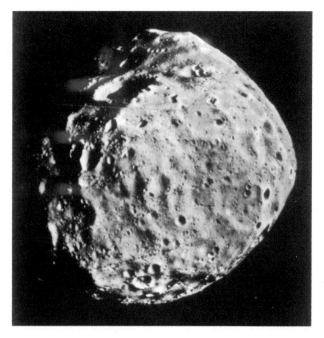

Figure 13. A captured asteroid. Phobos, 26 kilometres in diameter, is the larger satellite of Mars and a look-alike for a planetesimal. (NASA Reference Publication 1109, 1984.)

This is because they are consumed over time in the nuclear furnace in the Sun.

In the current understanding, planetesimals form after the original mass of gas and dust has formed a rotating disk, the stage at which it is most commonly thought of as the solar nebula. In a turbulent disk, bodies may grow to a kilometre in size within ten thousand years. This is the material that survives the early violent events in the inner solar nebula. These survivors eventually accumulate into larger bodies that ultimately collected together to form the inner planets and the Earth that we find so useful to live on.

Formation of the planets

Although we see many differences among the Earth and its relatives among the inner planets, one might have thought that the giant planets at least would be uniform and similar in composition to the Sun. Such a model would simplify our thinking about how to build planets. However, as we gained more information about the giant planets, they all turned out, like their terrestrial relatives, to be just as complicated as the members of any family group.

Most of the gas in the nebula was swept away within a few million years of the formation of the Sun. Since the gas giants had to form before the gas departed, an early formation date for Jupiter was inevitable. Saturn also has to have formed rather quickly, but at a slightly later stage than Jupiter, when less of the gaseous nebula remained. Uranus and Neptune must have accumulated at a later time still, for they are mostly ice and rock, with only a little gas.

The fundamental problem in forming Jupiter and Saturn is the need to explain the formation of rock and ice cores ten times or so the mass of the Earth so far from the Sun at a very early stage. This seems strange since the nebula became less dense with distance away from the Sun. Why didn't Jupiter form closer in? These problems will become clear when I talk about the current model for forming Jupiter before the gaseous nebula had totally departed.

The inner rocky planets formed from the left-over rocky debris in the inner solar system after the early violent Sun had swept away the gas and volatile elements. At that time a whole crowd of planetesimals ranging from metres to thousands of kilometres in size was milling about in the inner solar system. They slowly collected into four dominant centres.

What did these now vanished bodies look like before they came to a violent end as they were swept up into Earth or Venus? Were they already miniature planets, with separate rocky mantles and iron cores? For the bigger ones at least, the answer is yes. Probably, they resembled the asteroids, such as 243 Ida. Other look-alikes are Phobos and Deimos, the satellites of Mars, which are almost certainly captured asteroids.

How many planetesimals were needed to form the inner planets Mercury, Venus, Earth and Mars? Computer simulations indicate that just before the final sweep-up, there were over a hundred objects the size of the Moon, ten with masses exceeding that of Mercury and several exceeding the mass of Mars. The twins, Venus and the Earth, acquired most of them (Mars is only about one tenth of the mass of the Earth, while Mercury is only about one twentieth). Some of the largest bodies, the size of Mars, would have made respectable inner planets in their own right if their fate had taken a different course.

How long did it take to put the Earth together? The inner planets formed in a much more leisurely manner than the giants. The gas was long gone. In the depleted inner reaches of the nebula, it took somewhere between 10 and 50 million years for the scattered survivors to finally collect themselves together. Finally, they assembled themselves into the two bigger and two smaller planets that are so familiar to us. The largest collisions were among the last events to take place. The catastrophic crashing of planetesimals into a growing planet reminds one of the common experience of soldiers during warfare; long periods of boredom punctuated by brief periods of extreme terror.

The size of the solar system

An interesting question is the size of the solar system. Does this have any special significance? Was the solar system much bigger in the past? The solar system itself extends out to the fuzzy edge of the Oort Cloud of comets. The presence of another large planet would make for a nice round number of nine planets and perhaps fulfil some mystical numerological purpose. But then you might want to include Pluto or Ceres, and what about Ganymede, Callisto, Titan and Triton? Trying to make an exclusive definition of a planet is about as frustrating as trying to reach a satisfactory definition of life. How does one include mules but exclude automobiles in such attempts. One quickly becomes bogged in a semantic swamp.

However, of more direct interest to us is whether there are any large planets beyond Neptune? Were there more big planets out there, as Kant imagined? Being so far away, Neptune is only weakly bound to the Sun. Its orbit would easily be altered by the passage of a passing star, even one only ten per cent of the mass of the Sun. There is no sign of such an effect. This beautiful blue giant is in an extremely circular orbit that differs by only a very small fraction from a perfect circle. This is much better, as Bill Kaula (b. 1926), a geophysicist from the University of California, has remarked, than most machine shop specifications.

A large planet farther out would easily alter the orbit. Since this has not happened, Neptune is the true outermost planet. If large planets further out in the solar system had been stripped away by passing stars such events would have changed the orbit of Neptune. Clearly this has not happened. The planet continues on its stately course as it has done for the past several billion years.

Planet X has been a prime candidate for a major outer planet but direct searches for it have been unsuccessful. No gravitational effects due to the presence of such a planet have been noted affecting the orbits of the *Pioneer* or *Voyager* spacecraft, now in the distant reaches of the solar system. Surveys by satellites in the infra-red part of the spectrum have not detected any sign of the postulated 'Planet X', nor is there any sign of a dark companion star to the Sun, nor that other mythical beast, Nemesis, the Death Star, whose task is to send swarms of comets to devastate the inner solar system. Such periodic catastrophes have been held responsible for an alleged 26 million year repetition of extinctions found in the fossil record. As I discuss later, neither this periodicity, nor a supposed correlation with impact craters on the Earth, have survived a careful study.

So the expectation of Kant that many outer planets existed, complete with inhabitants, has not been confirmed. The orbit of Neptune is the true outer boundary of the planets.

The long-term stability of the solar system

Amongst other questions, we need to enquire whether the present arrangement of the solar system is stable. As computers have increased in power, they have been used to study the problem. However, definite proof of the stability of the system has not yet been achieved. The good news is that although the planetary orbits show some changes, these are small and the orbits are stable probably over the age of the system, as Laplace had calculated.

The chief argument, in the absence of definitive proof for the overall stability of the system, lies in its great age. Geologists and geochemists take comfort from the long-term stability that is obvious from the geological record. Sedimentary rocks that were laid down by running water are found back to the limits of the geological record, nearly four billion years ago. Their presence in that ancient epoch, and throughout the subsequent ages, shows that temperatures on the surface have been mostly between the freezing and boiling points of water. These even temperatures have been maintained despite the 'faint early Sun' problem that I will discuss later.

Many orbits in the asteroid belt are stable. Its zoned nature has persisted since the earliest times. Very little mixing among the different meteorite classes seems to have occurred over billions of years. The large asteroid Vesta seems to have been sitting in a stable position for over four and a half billion years, although at some stage chunks were knocked off it in a massive collision. These drifted into a site where the influence of Jupiter tossed some of them into Earth-crossing orbits, so providing us with the basaltic meteorites that we now call eucrites.

Most other orbits between the planets become unstable. Bodies in these gaps are soon swept up by the planets. Any comets that happen to wander into the gaps between the planets, as Chiron and the wandering herd of centaurs have done, are in for a short life.

2

The giants

Most of the matter in the solar system resides in the Sun and Jupiter. The rest of the system is so small that it can be ignored in a first approximation. Why is Jupiter the dominant planet? How did the solar system come to be populated by giant planets that reside so far from the Sun? How did they arise? Why are they not closer in to the Sun and why are there two sorts of giants?

The yellow and orange gas giants

Early perceptions

Jupiter and Saturn are splendid planets. With their associated rings and satellites, they excite the admiration of all observers. One perceptive writer on the space programme compared the giant planets to French impressionist paintings: 'Jupiter's orange and yellow bands are so roiled up that its disk might have been painted by van Gogh at Arles. Saturn, with more delicate bands of ochre, resembles a Monet haystack in a sunlit mist. Uranus' disk, though, is so featureless and limpid as to suggest the still pond around a Monet water lily'.[1]

The ancient astronomers, unaware of the true size of Jupiter, nevertheless showed unusual insight in naming it after the chief Roman god, for we might not be here if Jupiter did not exist. First, it cleaned out the solar system, and now it acts as a shield against the impacts of comets. Without the mighty gravitational shield of Jupiter, the bombardment of the Earth today would rival that of warfare. Saturn, named after the Roman god responsible for agriculture, was recorded in the seventh century BC by the Babylonian astronomers. It was the outermost planet known until the discovery of Uranus in 1781, followed by that of Neptune in 1846.

Difficulties in forming gas giants

Jupiter presents us with several major problems. The rocky terrestrial planets, including the Earth, which was once regarded as the centre of the universe, are trivial in comparison. They formed much later than Jupiter, from the rocky bits and pieces left after the gas and volatile elements in the inner solar system had been swept away.

However, Jupiter had to form while the gas is still around. For other reasons, it is also clear that Jupiter formed very early. It had to have formed before the asteroids. There is a great 'hole' in the solar system at the asteroid belt. There are many thousands of asteroids out there but their total mass is trivial, even compared with that of the Moon. When they are all added together, the mass reaches about five per cent of the mass of our satellite. The material that was originally present in the disk at that location has nearly all gone. As no one has proposed that there was a hole in the original disk of dust and gas in that region, the general view is that it is the fault of Jupiter. Once this giant grew big enough, it grabbed everything within reach. It's a good example of the rule that the rich get richer. What this massive planet didn't take, it tossed out to the outer reaches of the system or out of the solar system entirely, behaving like a dominating and careless giant.

Jupiter caused a second, but equally devastating effect among the surviving planetesimals in the asteroid belt. Its strong gravitational effect distorted their orbits so that they were unable to collect themselves into a planet. They still wander around like a crowd of refugees, denied the opportunity to settle down to make even a tiny planet.

Mars suffered from a slightly different problem. It grew up in an impoverished neighbourhood, and its growth was stunted, having to put itself together with what pieces were left by Jupiter. Perhaps as much as 97 per cent of the material which Mars could have used, had been taken by the giant. Mars is thus a tiny planet, only a bit over ten per cent of the mass of the Earth.

So the argument for early growth of Jupiter depends on several observations. These include the large amount of gas in Jupiter, the absence of a planet in the asteroid belt, the small size of Mars, and also the small number of planets. All appear to require the early formation of a dominating giant that contains most of the mass of the disk.

How to build Jupiter early enough: blizzards in the nebula

If that was the whole story, it would be simple enough to condense Jupiter early from the nebula, like a second small sun or brown dwarf. That was once a popular way to make planets, as I discussed earlier. Jupiter, however, is not a failed star, but a true planet, built up bit by bit from the nebula. If it was a condensed bit of the original nebula, then it would have the composition of the Sun. Instead, it has less than a tenth of the gas expected. It also has too much ice and rock and too little gas compared with what was available in the early nebula.

As I discussed earlier, much of the rock and ice is concentrated deep inside the planet in a core. Once such a massive core, mostly of rock and ice and perhaps ten times more massive than the Earth, has formed, gas from the nebula will collapse on to it by gravitational attraction. In this fashion, a great gas giant can form, in this case with over 300 times the mass of the Earth (see Figure 14).

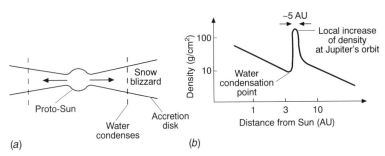

Figure 14. As the early Sun turned on its nuclear furnace, strong stellar winds drove water and other volatile material out to about 5 AU where the water condensed as ice and piled up in a snow line (*a*). This increased the density of the nebula at this location (*b*) and so enabled an icy core about ten times the mass of the Earth to form quickly . This core could then capture some of the gas that was also being driven away from the violent early Sun. The result was the rapid growth of the planet Jupiter. Adapted from Stevenson D. J. (1989) in *The Formation and Evolution of Planetary Systems* (H. A. Weaver and L. Danly, eds.) p. 85. Cambridge University Press, New York.

But there is a more serious problem in the case of Jupiter. How does one form such a big planet so far from the Sun. Out at 5 AU the density of the disk of dust and gas was low. And how can all this happen so early in the history of the solar system? Such a massive core has to form very quickly while the gas is still around. The gas in the nebula lasts for only a few million years before it is swept away. So Jupiter has to have formed a core quickly enough to catch the gas. If the core was a little late in forming, or grew too slowly, the gas would have already gone and we would see a planet of rock and ice. There would be an icy giant like Neptune or Uranus in place of our great Jupiter. George Wetherill (b. 1925) at the Carnegie Institution of Washington, impressed by the difficulty of constructing such a giant planet, has commented that 'I don't think the existence of Jupiter would be predicted if it weren't observed'.[2] So we are very lucky to have this giant planet, whose gravitational shield protects the inner solar system.

Coming back to the problem of assembling Jupiter, the prime need is to form a large core of rock and ice out at 5 AU from the Sun. Why there? It would make more sense to form it closer to the Sun, where the nebula was denser. As we saw earlier, the original disk of dust and gas was made up of three main components, gas, ice and rock. As the Sun lit up, intense solar winds drove the gas and ices away within a few million years. However, out at about 5 AU, the future site of Jupiter, it was cold enough that the water being swept away froze to ice. There it piled up at a sort of 'snowline', leading to a massive build up of ice. This led to a great increase in density and the core of Jupiter grew rapidly as all this extra material arrived. This embryo grew to something like ten times the size of the Earth in as little as half a million years. When it got big enough, it started to capture some of the fleeing gas. Jupiter was able to snatch enough to grow to become a giant, Even so, Jupiter only collected ten per cent of the original gas, and ninety per cent escaped.

The rest was swept away by the early intense solar winds. Left behind in our region in the inner nebula was a population of boulders and rubble big enough to resist being carried along in the gale. We are standing on a rock pile that was assembled later from the survivors.

The limits to growth

Why did Jupiter stop growing? What were the limits to the growth of the giant planets? Why are they not larger? Why did not all the material in the nebula finish up in one giant planet, so that the system would resemble a double star? The answer is that Jupiter, having taken everything within reach, cleared a gap in the nebula and ran out of material within its grasp.

For the ice giants, Uranus and Neptune, the answer is also clear. In the less dense regions of the outer solar system, these planets took longer to form a core. They formed too late to catch much gas before it was all gone. Thus, the formation of planets is self-limiting; they run out of material. In a bigger nebula, space travellers would have seen yet another pair of double stars, instead of a single star surrounded by eight planets and some odds and ends. They perhaps would have noted its unique character for they were just as likely to have seen other planetary systems that were wildly different from our own.

Some internal problems

The atmospheres of Jupiter and Saturn, with their variety of colours, provide some intriguing problems. The famous red spot has been observed through telescopes for 300 years, and marks the site of a great cyclonic storm system. We still don't know what the colour is due to. It used to be believed that the compositions of the atmospheres of Jupiter and Saturn were the same as that of the Sun. This would fit in with the idea that they were unaltered fragments of the primitive disk of gas and dust. Even though we now know that these planets have less gas than was there to begin with, one might have expected that the ratio of hydrogen to helium would be the same in the planets as that in the Sun. Everyone knows how difficult it is to separate two well mixed gases. It takes sophisticated equipment on Earth to separate oxygen from nitrogen. So it was surprising to discover that measurements of the two gases in the atmospheres of Jupiter and Saturn showed large differences from the solar ratio. Jupiter seemed to have lost some helium while the atmosphere of

Saturn is even more depleted, containing only about a quarter of the solar abundance of that gas.

Both Jupiter and Saturn are so large that the helium could not escape the gravitational grasp of these giants. And after all, it is the lighter gas, hydrogen, that should have escaped first. Thus, it was realised that this loss must be due to some process within the planets.

The answer turns out to be rather simple. At low temperatures, as the planets cooled, liquid droplets of helium formed like rain drops condensing from water vapour. As they are dense, they fell, just like rain, toward the centre of the planet. Hydrogen, the other main component, remained as a gas. This caused helium to concentrate toward the centre of the planet. The atmosphere became richer in hydrogen. Saturn is smaller than Jupiter and has cooled more quickly. Just as in any cold climate, there has been more rain. So helium is scarcer in the atmosphere of Saturn than in that of Jupiter.

Are Jupiters common or useful?

Finally, it is worth addressing the question as to whether planets like Jupiter might be common in other planetary systems. If the formation of large gas giants is part of the same lottery of chances that is such a feature in our solar system, Jupiters might be uncommon. It certainly seems to have been difficult enough to form Jupiter. The timing is exquisite. If the Sun had been bigger, or had a more violent early history, the gas might be swept away before a core big enough to catch it could form. Then one would be left with icy giants, clones of Uranus and Neptune. Suppose that the core forms just too late to catch the gas. We finish up again with a Uranus or Neptune in place of Jupiter. Suppose the 'snowline' didn't form. Then we might have a much smaller size planet at Jupiter, and a larger version of Mars, perhaps half the size of the Earth. There would be a respectable planet in place of the asteroid belt. Probably all these processes occurred, as well as some that we haven't thought of, around other stars out in the galaxy. This view is reinforced by the variety of new 'planetary' systems. They fit no simple pattern. One possibility is that more than two gas giants form. Model calculations for this scenario look like the result of a crazy baseball game, as the tidal interactions toss the giants around. More on this later.

The Roman perception of the importance of Jupiter has a modern analogue. This is mainly because Jupiter cleaned out the neighbourhood, throwing away the material that it did not need. Apart from this useful cleanup, Jupiter acts as a gravitational shield, collecting comets that stray into the inner solar system. Thus, if Jupiter did not exist, or was smaller, the Earth would be bombarded with comets. The numbers of impacts on the Earth would be perhaps a thousand times greater than at present. Local areas, a few kilometres across, would experience a catastrophe several times a year, instead of once in a thousand years. Collisions of the sort that killed off the dinosaurs might occur every hundred thousand years rather than every several hundred million years or so. This bombardment would have had incalculable effects on the development of life, perhaps stopping it altogether. Even with the protection of Jupiter, life on this planet has come perilously close to being extinguished. It seems unlikely that our species, or life itself could have survived such disasters without the protective shield of Jupiter. In the rather brief time span that the genus *Homo* has been around, we would have experienced between 20 and 100 such interesting impacts. Perhaps we would all have retreated far underground, and so would never again contemplate the starry heavens.

The green and blue ice giants

The triumph of the Newtonian system

Uranus was the first planet to be discovered since antiquity, although its discovery by William Herschel (1738–1822) in 1781 was accidental. He called it George's Star (Georgium Sidus) after King George III, who then presented Herschel with a lifetime pension. Such nationalistic fervour, although rewarding for Herschel, did not receive universal acclaim. The planet was eventually given the appropriate classical name of Uranus. However, it was soon discovered that the orbit of Uranus was being affected by another large body further away from the Sun. Study of these variations in the orbit of Uranus led eventually to the discovery, in 1846, of the cause. This was the existence further away from the Sun of another large planet, Neptune.

The discovery of Neptune was not without irony and forms a cautionary tale. J. C. Adams (1819–1892) in England, and Urbain Le Verrier (1817–1877) in France independently calculated where a planet might be found on the basis of its effect on the orbit of Uranus. By September, 1845, Adams had worked out where to look, but there was bureaucratic delay in England in acting on his predictions. Meanwhile Le Verrier was having similar difficulties with his French colleagues, and finally persuaded the astronomers at the Berlin Observatory to search. The planet was discovered by J. G. Galle (1812–1910) during his first attempt on September 23, 1846. It turned out that the observers in England had seen but not recognised the planet, several weeks earlier. Galileo probably also saw Neptune over 200 years earlier, for it appears on one of his star charts, but he did not recognise that it was a planet.

These discoveries had a profound influence on western thought. It was a dramatic demonstration that the natural laws discovered by Isaac Newton had the power to make precise predictions of the motions of the planets. The universe indeed appeared as well ordered as a clock, behind which perhaps was the clockmaker.

Differences among the giants

Although the solar system is often divided into the inner rocky and the outer gaseous planets, Uranus and Neptune are very different from Jupiter and Saturn. The ice giants are dwarfs compared with Jupiter, which is 318 times more massive than the Earth. Uranus is only about 14 times more massive the the Earth. Curiously enough, Neptune, although one third further away from the Sun than Uranus, is about 20 per cent more massive than that planet. Both these ice giants are composed of a mixture of ice and rock, and contain only a little gas. In this way they resemble the core of ice and rock around which Jupiter grew. Jupiter is basically like Uranus or Neptune with the addition of a thick shroud of gas. Neptune being further away from the Sun than Uranus, finished up with a smaller amount of gas. Although a simple theory would predict that planets should get less dense with distance from the Sun, Neptune in contrast is also denser than Uranus. This higher density is due to Neptune having less gas and more ice and rock than its inner neighbour, so that pressures (and density) are higher inside the planet. For this reason, Neptune, although more massive, is smaller than the more bloated Uranus, which has swollen out from containing a little more gas.

Origin of the ice giants

Why is there so little gas in either planet compared with the abundance in the gas giants, Jupiter and Saturn? Both Uranus and Neptune managed to form massive cores, by accumulating a myriad of icy and rocky bodies of which Pluto, Triton, and the Centaurs, which I discuss a little later, are surviving examples. So far away from the Sun, the outer reaches of the dusty disk were thinly populated. It took perhaps ten million years before Uranus and Neptune had grown big enough to capture the gas being swept away from the violent Sun. By the time their cores got big enough to capture it, the gas had mostly fled. Uranus and Neptune thus suffered the usual fate of coming too late to the table. It is for this reason that these planets are small icy giants compared with the massive gas giants, Jupiter and Saturn. If the original disk of gas and dust had been larger, they might have grown to rival Jupiter or Saturn. In a smaller disk, they may have become the largest planets.

However, we have only very general models for building these planets. We understand why the gas content falls off from Jupiter to Neptune, but what triggered the runaway growth of the cores? This is understood for Jupiter due to the pileup of ice at the snowline. But perhaps more rock piled up there and more ice at the site of Saturn, or two cores were formed and one migrated outwards. Maybe similar condensations of ammonia and methane ices were responsible for the rapid growth of cores further out in the colder reaches of the nebula. What is clear is that simple banging together of planetesimals to construct planets takes too long in this remote outer part of the solar system. The time needed exceeds the age of the solar system. We see Uranus and Neptune, but the modest requirement that these planets exist has not been met by this model.

The difference inside

In models for the internal structure of Uranus and Neptune these planets have a rock and ice core. This is covered with an icy oceanic shell, which is a mixture of water ice, methane and ammonia. Then there is an outer envelope composed of gas and some ices. However, these layers are not clear-cut and they grade into one another.

However, this is not the end of the story for these apparent twins. Being so much further away from the Sun, Neptune receives less than half of the sunlight that strikes Uranus. One might expect that Neptune would be colder than Uranus. However, curiously enough, their surface temperatures are identical at 59 degrees above absolute zero. Neptune seems to have made up for its remoteness from the Sun by retaining a substantial amount of internal heat. This is probably left over from the energy associated with assembling the planet. Uranus, in contrast, has a very low, and possibly zero, flow of heat from its interior. Perhaps it lost all of its original store of heat. More likely, it is still bottled up inside. Thus, the interiors of these two planets must be different to account for the difference in the temperatures inside them. These differences are not related to distance from the Sun, and present a good example of the sort of detail that needs to be accounted for in theories of the formation of the solar system. It is ironic that the one property that is identical, the surface temperature of these two icy giants, turns out to be due to different causes. The devil is in the details.

This pair of icy giants once again demonstrates the difficulties in making similar planets, even in the cool outer parts of the solar system, far from the early violent Sun. One might have expected that things might become more uniform on approaching the outer edge of the system, but that hope is not realised.

Tilts and giant collisions

Uranus and Neptune rotate at about the same speed, resulting in days of 17 and 16 hours respectively. However, they have dramatically different tilts. The axis of rotation of Neptune is tilted, like Saturn, at nearly 30 degrees to the common plane of the solar system. Uranus, in contrast, is lying on its side (see Figure 15). Its satellites

Figure 15. The planet Uranus, lying on its side, with its set of nine dark rings orbiting its equator.

and nine dark rings rotate about the equator, an interesting problem to be discussed later. The only way to knock Uranus over was to hit it with something big. Whatever did the job had to be about the mass of the Earth. Models for forming the planets also form a variety of smaller bodies, up to Earth size, that eventually collide with a planet. It has been suggested that the difference in the heat inside Neptune and Uranus may be due to different internal structures resulting from this massive collision, just as a boxer might have had his internal organs rearranged from a great blow in the solar plexus.

The outer edge of the planets

As I discussed earlier, Neptune is the true outer boundary of the planetary system, which ends rather abruptly at that planet. No large or even small planets exist beyond Neptune. Tiny Pluto is an interloper that I talk about later. Neptune had been discovered because it caused wobbles in the orbit of Uranus. For a long time, there were also thought to be variations in the orbit of Neptune. These deviations encouraged the speculation that there must be yet another major planet lurking out there in the far reaches of the system. Pluto was of course much too small to be a candidate. Thus, the idea of the dark planet X arose. Human imagination likes to create hidden monsters.

Alas, the calculated variations in the orbit of Neptune have turned out to be spurious. They were due due to an incorrect value (by one part in 200) of the estimate of the mass of that planet. When the correct mass, obtained from the gravitational effect on spacecraft passing near the planet, is used in the calculations, the calculated wobbles in the orbit of Neptune disappear. What is out beyond Neptune is the source of comets from which we receive an occasional icy visitor.

Satellites of the giant planets

Miniature solar systems?

Circling around Jupiter are the four famous Galilean satellites, Io, Europa, Ganymede and Callisto. When Galileo saw in 1610 that these bodies were in orbit around Jupiter, he realised that this provided direct proof of the correctness of the Copernican System. They resemble a miniature solar system, a point also appreciated by Galileo. Their orbits around the equator of Jupiter are nearly circular. These well behaved satellites are fairly uniform in size and they show a steady decrease in density with distance from the planet. This is because the amount of rock falls off and that of ice increases in the satellites as one goes away from Jupiter. All these regular features encouraged the view that the Galilean satellites would provide insights into the formation of the solar system, just as one can study the workings of a large machine from a scale model.

Perhaps there were further insights to be gained from looking at the satellite systems. We have only one solar system, and nothing similar with which to compare it, not even the new 'planetary' systems discovered so far. The statistics of one bedevil attempts to understand the solar system, but there are several sets of satellites. Perhaps a study of these could provide some universal rules for building solar systems, just as a blueprint or a computer program enables one to construct large machines. There are a host of satellites around the four giant planets. Three of the giant planets possess regular satellite systems that mimic the solar system in miniature. Neptune has yet another set of satellites, dominated by Triton, of which I shall talk about later. These four scale models provide us with some statistics. Surely we might gain some insights into the formation of planets from studying them. This hope is soon dashed. The satellite systems of the four giant planets are startlingly diverse. Their satellites are so distinct that they could all belong to different planetary systems without causing any surprise. To add to the confusion, the satellites of Earth and Mars are exceptional cases, even by the rather broad standards of the solar system, and are of little use in this enquiry.

The startling diversity

There are 60 or so satellites (Figure 16), but no two are truly alike. Various attempts have been made to classify satellites, but like many features of the solar system, they mostly defy being placed into neat pigeonholes. However, their names are so enchanting that I make no apology for listing many of them.

The first attempt at a classification puts them into three divisions: regular satellites, irregular satellites, and a third category of bits and pieces resulting from collisions. The regular satellites are in prograde orbit around the parent planet, that is, they orbit their parent planet in the same sense that the planets go around the Sun. They are generally thought to have formed from disks around the parent planet. These satellites include almost all the larger ones. These include the four Galilean satellites of Jupiter (Io, Europa, Ganymede and Callisto), the eight classical satellites of Saturn (Mimas, Enceladus,

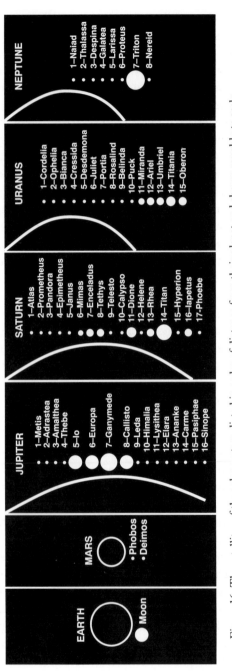

Figure 16. The satellites of the solar system, listed in order of distance from their planet and shown roughly to scale.

Tethys, Dione, Rhea, Titan, Hyperion and Iapetus) and the five classical satellites of Uranus (Miranda, Ariel, Umbriel, Titania and Oberon).

The irregular satellites mostly have highly inclined and elliptical orbits far away from the planet. They include the four members of the cluster (Leda, Himalia, Lysithea and Elara) that are in a prograde orbit around Jupiter. Another group (Anake, Carme, Pasiphae and Sinope) are quite remote from Jupiter and all rotate backwards around the planet. Other irregular satellites include Phoebe, the outermost satellite of Saturn and Nereid, the outer satellite of Neptune. Most of these distant satellites are probably stray comets or planetesimals that have been captured. The chief interest in these captured bodies is that they are probably left-over building blocks of the planets. They survived being swept up into the planets. Like fugitives from a lost battle, they were rounded up and captured late in the day.

Most of the remaining satellites are, in the words of one observer 'tiny, craggy chunks, battered and worn down by the ongoing meteoroid flux'.[3] Examples include Metis, Adrastea, Amalthea and Thebe, which are all embedded in Jupiter's ring. Around Saturn, there is Atlas (skirting the outer edge of the main A ring), Prometheus and Pandora (the shepherds of the F ring), Janus, Epimetheus, as well as Helene, Telesto and Calypso (trailing the orbits of Tethys and Dione). In addition there are ten small inner satellites of Uranus that were discovered by the *Voyager* spacecraft in 1985–1986. Finally, for anyone who is not totally lost in the outer solar system by this point, there are the inner satellites of Neptune. These are, in order outwards from the planet, Naiad, Thalassa, Despina, Galatea, Larissa and Proteus. They are bits of rubble, probably the remnants of some larger satellites that were destroyed during the capture of Triton, which I talk about later. These fragments fit about as well in this category as anywhere.

Some objects refuse to be put into even these three broad pigeonholes. These include Triton, Charon, the Moon, and the tiny satellites of Mars, Phobos and Deimos. Triton revolves backwards around Neptune and is a close relative of Pluto. Charon, the moon of Pluto, probably formed when something collided with that icy body. Both Triton and Charon are special enough to demand separate treatment. It is also ironic that our nearest neighbour, the Moon, cannot be placed into any tidy organisational scheme, and likewise needs a separate chapter to explain how it came to be in our night sky. Our 'twin' planet, Venus, adds to the problem by not having a satellite at all.

Finally, the tiny satellites of Mars, Phobos and Deimos, are often placed in the category of captured objects. This is because they are different in composition to Mars. Once they were captured, these small bodies were quickly forced into circular orbits by Mars. Phobos, about 26 kilometres on its long axis, is slowly spiralling inwards and will hit Mars within the next 40 million years. A large basin the size of Belgium, surrounded by rings of mountains, will form on Mars as a result. Possibly this collision will throw some more meteorites to the Earth. It is a matter of regret that this spectacular and informative event will occur so far in the future on human time scales. One of my colleagues, an expert in impact craters, has suggested that we should nudge Phobos into a collision course with Mars. By this dramatic means, we could observe the formation of a impact crater over 200 kilometres in diameter 'in real time'.

It is clear that the search for some kind of regularities among the satellite systems has failed and that no simple sequence of reproducible events has occurred in the solar system.

The Galilean satellites of Jupiter

These four famous satellites, Io, Europa, Ganymede and Callisto, have four totally different surfaces and interiors. Galileo would surely have been pleased that his galilean satellites continue to provide new information 400 years after he first saw them.

Volcanic activity is widespread on Io, due to heating from tidal stresses from nearby Jupiter. Io has many hundreds of volcanoes, some up to 200 kilometres in diameter. The spectacular volcanic plumes seen by spacecraft are 150–550 kilometres in diameter and up to 300 kilometres high. These dwarf our volcanic eruptions. It is clear that the volcanic plumes may last for years. Eight of the plumes were observed by both *Voyager* missions, which passed Io four months apart in 1979.

The best match for the overall colour of Io is yellow, with orange, green and grey tones that make Io look like an over-ripe orange. These colours are due to various forms of the element sulphur, with overtones from other rocks and trace minerals. Although in popular belief Io is covered with lava flows composed of sulphur, in fact sulphur forms only a thin coating. The satellite has quite a rugged relief, with mountains up to 10 kilometres high. These are impressive for a body about the size of the Moon. Some volcanic craters are two kilometres deep. Only rock has the strength to support such structures, sulphur being much too weak. There are almost no impact craters. Most have been covered by lavas. Io has clearly melted early in its history for it has formed a large iron core, which now generates a magnetic field.

Europa, the next satellite out from Jupiter, has a frozen icy surface (see Figure 17). There are only a few impact craters showing on the surface. Europa, although further away from the giant than Io, is close enough to Jupiter to be warmed a little by mild tidal stresses, so that impact craters formed on the ice are soon wiped out. There are lots of fractures, just like those caused by jostling ice floes on Earth. The

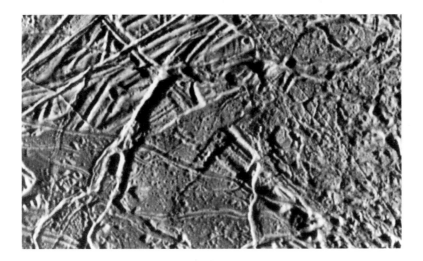

Figure 17. The icy surface of Europa, broken up into many cross-cutting ridges and grooves. This view is of an area only 10 by 16 kilometres recorded by the *Galileo* spacecraft when it passed within 3340 kilometres of the satellite in January, 1997. This picture forms an excellent example of the sort of detail now available to students of such remote landscapes (NASA P-48227).

surface looks a lot like that of our polar seas, with a jumble of ice flows and icebergs. Bits of the icy crust are upturned and rotated. This resemblance to the structures seen on our pack-ice tell us that the icy crust is probably only a few kilometres thick, and floating on an ocean. This scene resembles that of the Arctic Ocean on the Earth, except that the ocean on Europa is 200 kilometres deep. Further down there is a rocky mantle and a small iron core. Europa is thus like a slightly smaller version of our Moon, but with the addition of a couple of hundred kilometres of water.

The presence of an ocean raises the question of life. At some stage in the past, submarine eruptions of lava were very likely. These could give rise to hot springs like those at our midocean ridges. As these are possible places where life originated on Earth, this raises the question whether anything is lurking under the ice on Europa. Some europan relative of the Loch Ness Monster has probably already arisen in the media.

Ganymede is the largest satellite in the solar system. Although smaller than Mars, it is bigger than Mercury. It would not look out of place among the inner planets. Ganymede has a core, either of metal or iron sulphide, surrounded by a rocky mantle. Above this is an 800 kilometres thick mantle of ice. If one added such a sheath of ice to Io, it would closely resemble Ganymede. Just like the Earth, satellites have melted, forming cores and rocky mantles. There is no mystery how Io acquired the necessary energy – from tidal interactions with Jupiter. Ganymede is much further away. One possibility is that Ganymede became trapped in a tidal resonance and heated up, perhaps a billion years after it formed. This late heating event could explain the unique landscapes that we observe.

The most distinctive feature of Ganymede is the existence of two types of crusts. Each covers about half of the surface area of the satel-lite. The older darker crust is more heavily cratered and appears to have been split apart. Younger crust has formed between the cracks. The only reasonable explanation is that the satellite has expanded by a kilometre or two. This seems to have been due to melting of dense ice deep inside. As the water flooded out, it froze in our familiar less dense form of ice. Ganymede, the largest satellite in the solar system, is thus of extraordinary interest. It displays evidence of slight expansion which can be understood from the known physical properties of ice.

Callisto, the outermost of the Galilean satellites, forms a striking contrast to Ganymede. Callisto has an icy crust, covered with craters. It has not changed much since the end of the massive bombardment of the solar system and so it has recorded impacts for four and a half billion years on its permanently frozen crust. It is an undifferentiated body, a mixture of four tenths ice and six tenths of rock.

Why is Callisto a primitive body unchanged for aeons, while Ganymede has had a complex history? Clearly, Ganymede has been on one side of a critical boundary and Callisto was on the other. The most likely possibility is that Callisto is far enough away from Jupiter that it has escaped any tidal heating. Perhaps a less likely cause is that Ganymede is a little larger and a little denser than Callisto. Thus it may have generated more internal heat, leading to melting deep within the satellite. Callisto was just a little smaller and so stayed frozen throughout.The great contrast between the two satellites brings to mind the difference between Venus and the Earth, two bodies of similar size that differ so greatly in detail.

So even in this best behaved example of a satellite system, the four individuals don't resemble one another. Like almost all other bodies in the solar system, the Galilean satellites have unique features. This system provides yet another example of the amount of complexity even in this apparently simple set of satellites.

The satellites of Saturn

Laplace was aware that Saturn had seven satellites. Now we have found nine additional ones and there must be some smaller ones yet to be found in amongst the rings. In contrast to the relatively orderly Galilean system, the satellites around Saturn show few regularities. Those seeking some order may take comfort from the fact that the satellites do orbit Saturn in the same sense as the orbits of the Galilean satellites around Jupiter. However, in contrast to the relatively well behaved four that Galileo saw at Jupiter, they do not show any regular variation in density with distance from the planet. They are also less dense. There is only one large one, Titan. This satellite is well away from the planet, with several much smaller satellites closer in. Thus, the set of satellites around Saturn bears little resemblance to those around Jupiter. This is not encouraging for people who wish to produce planetary systems by tidy computer programs.

Several of the satellites are worthy of special comment, mainly to illustrate the ability of the solar system to produce unexpected forms. Thus, Enceladus has had a complex history. Parts of the surface are covered with craters. Other areas have few craters and so are younger. These regions have been resurfaced by water ice, probably a result of heating due to tidal stresses. Other satellites of Saturn are covered with impact craters, and so have changed little over the last four billion years (see Figure 18).

Hyperion, lying between Titan and Iapetus, is the fifteenth in order outwards from Saturn. It is tumbling chaotically and has a very eccentric orbit. Hyperion appears to be the remains of a satellite broken up by a massive collision. Calculations show that it will continue to tumble until the 'end of the world' some five billion years in the future. At that time the planets and their companions will be overwhelmed, when the Sun swells up in the red giant stage of solar evolution.

Iapetus adds yet another strange creature to the zoo. Like a circus clown, one face is black and the other is white and about ten times brighter. This causes the satellite to become nearly invisible to earth-based telescopes when the dark face is turned toward the Earth. The strange appearance is due to debris from an impact on neighbouring Phoebe. This has coated the leading face of Iapetus as though it was the victim in a celestial pie-throwing contest.

Titan is unique in the solar system. This satellite has always excited interest, both on account of its size and for the presence of a thick atmosphere. It is only a little smaller than Ganymede. Like that satellite, it is bigger than the planet Mercury. It has an atmosphere mostly of nitrogen with some methane and a surface pressure of 50 per cent higher than that of the terrestrial atmosphere. The surface temperature, however, is only 95 degrees above absolute zero, cold enough so that the methane may be liquid. So there might be rivers, lakes or oceans of methane, under an orange sky. The atmosphere is a sort of a nightmarish version of smog, full of complex organic chemicals. Titan is a favourite target for searches for chemical precursors to life or possible life forms, but what sort of creatures might live in a lake of methane is difficult to imagine.

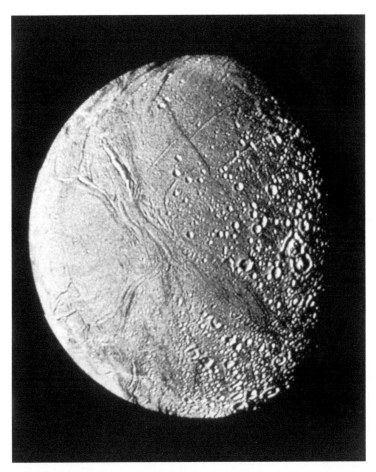

Figure 18. Enceladus, an icy satellite of Saturn, 500 kilometres in diameter, which displays a complex geological history (NASA *Voyager* P-24308).

The satellites of Uranus and Neptune

In 1767, Herschel, by using a large telescope, discovered Titania and Oberon moving round Uranus in a plane vertical to the rest of the solar system. Since then, we have discovered another 13, all orbiting around the equator of the giant. The satellites of Uranus have densities generally higher than those of Saturn. This is the reverse of what one might have expected for bodies further away from the Sun.

These small icy bodies might have been expected to be small versions of Callisto, frozen icy lumps covered with craters. Instead, Ariel and Miranda show surprisingly young surfaces. They have only a few craters. What seems to have happened on these frozen worlds is that mild warming has caused slushy mixtures of ammonia and water ice to flow around on the surface. These look like lava flows in the photos. The older impact craters have been covered over and the landscapes look like lava flows on the Earth. Like so much else in the outer solar system, this was unexpected.

Miranda, discovered by Gerard Kuiper (1905–1973) in 1948, is exceptional even by the rather broad standards of the solar system. It has a very large inclined and eccentric orbit and yet another extraordinary surface that outdoes that of Io or Ganymede (see Figure 19). Totally different landscapes occur side by side. Some areas have mountains up to twenty kilometres high. Nearby there are great canyons. It looks as though the satellite was broken up by a collision, and the bits and pieces just thrown back together anyhow. No one predicted anything looking like Miranda. It remains as an inspiring

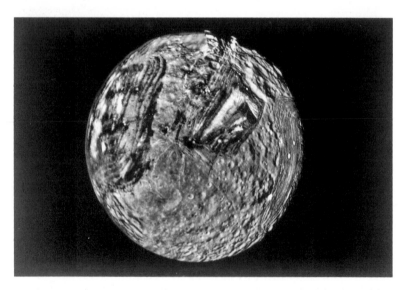

Figure 19. Was this satellite of Uranus broken up and reassembled? The bizarre surface of Miranda, 480 kilometres in diameter (NASA *Voyager* P-30230).

landscape at which to wonder about the great differences in the solar system. Nature has a surprising capacity to produce bizarre results within the confines of physical laws.

Neptune has eight satellites. Six are tiny icy chunks lying close in to the planet. Further out, Triton, which is bigger than Pluto, orbits Neptune backwards in a highly inclined orbit. It is a first cousin to Pluto and is probably a captured icy body. It is sufficiently important to need separate treatment, which will come later. Finally, about 15 times further away from Neptune, tiny Nereid occupies a wildly eccentric and inclined orbit.

The strange satellite system around Neptune seems to be mainly a ruin, the result of the capture of Triton. The arrival of this large intruder on the scene must have resembled that of a bull entering a china shop. Any original satellite system was destroyed by collisions on very short time scales. The tiny inner satellites that we now see may be survivors from the debris that remained. Five of them lie close to the rings that surround the planet and perhaps are even broken-up bits left over from the ring-forming event. Faraway Nereid is probably a captured comet.

The satellites of Neptune have a particular message. They show what can happen to a regular system of satellites if a large body is captured. One out of four satellite systems in our own solar system was disrupted. Although the capture of Triton was a random event, it shows how frequently such disasters could occur.

The cosmic junkyard

The satellite systems of Jupiter, Saturn and Uranus have so few features in common that they could belong to separate planetary systems. Thus, the satellite systems are all individual and specific to the planet around which they orbit. There is no grand programme for manufacturing them on some sort of cosmic assembly line. The solar system does not look like the product of a well organised factory for producing planets and satellites. Instead, it looks as though it was assembled from the bit and pieces lying around in some cosmic junkyard. This gives little cheer to the idea that close relatives of our solar system exist. If the four giant planets produce such different sets of satellites, we should not be surprised at the differences among other planetary systems that we now observe. Although other planetary systems may be common and life may have begun elsewhere, the message here is that nothing resembling the Earth, nor *Homo sapiens*, is likely to exist elsewhere in the universe, a theme that I elaborate on in later sections

How are satellites made?

The irregular satellites are easy to deal with. They are mostly captured bodies, of interest as left over bits that did not get collected by the planets as they grew. In contrast, the regular satellites are supposed to form like a miniature solar system from disks around planets. This raises some interesting questions. How do such disks form? Disks might form naturally around growing planets, or be spun off as the planet contracts. However, only Jupiter and Saturn may have formed satellites in this way, since these planets originated while the gaseous nebula was still present.

Disks could also form from material spun off or kicked off the planet by massive collisions. The satellites of Uranus and Neptune may have formed from such disks since the gas in the nebula was mostly gone by the time that they formed. In this regard, there is some insight to be gained from the satellites of Uranus. These all orbit in the plane of the equator and must have formed after the collision that knocked the planet over on its side. The gas blown off by the collision may have become a disk from which the regular satellites condensed. If the satellites had been there before the collision, it seems very unlikely that they would rearrange themselves through an angle of 90 degrees following the large impact. When this event happened, Uranus must have been close to its present size. In this scenario, satellites form later than the planets.

It will be apparent from this discussion, heavy with speculation, that our thinking about satellite formation is even less clear than that of the origin of the giant planets. Even if the formation of satellites is a common process, it does not seem to produce uniform results, even around such planets as Jupiter and Saturn. No single model can accommodate the complexities observed. Accordingly, we are faced with the same dilemma as with the planets. There is no grand unifying theory to account for all the different observations. The idea that some kind of universal process for constructing planetary systems is certainly not obvious. The bizarre new 'planetary' systems that are beginning to be discovered reinforce this notion.

Finally, it is worth noting that the spacing of the satellites about the giant planets, although regular, is a secondary effect. They are small and relatively much closer in to the primary planets than are the planets to the Sun. Their orbits are thus easily changed by tidal forces. Thus, their spacing is secondary and not a primary feature. Some models try to account for these spacings as due to initial condensation from the nebula. But the spacings are not original, and have evolved with time. Thus, they tell us nothing about the origin of the solar system.

3

Escapees and survivors

The formation of the giant planets affected everything else in the solar system, comets, asteroids, and the inner planets, to some degree. Some bits and pieces were scattered or escaped from the giants. Other bodies were survivors that assembled themselves out of what the giants left. These unfortunate creatures include the comets, asteroids and even the planet Mars. Although one might also include Mercury in this list, it has had such a traumatic history that, like a special patient, it needs separate treatment.

Comets

Ghostly apparitions

The appearance of comets in the night sky caused much distress in primitive societies, because the starry heavens were thought to be fixed and permanent. Often comets were thought to be some kind of disturbance in the atmosphere of the Earth. Even now the appearance of a comet is called an 'apparition', a word meaning in popular usage a ghostly figure or a sudden or unusual, frightening sight. The appearance of the ghost of Hamlet's father in Shakespeare's play meets this definition. Even when not terrifying people, comets were thought to herald great changes. Shakespeare tells us in *Julius Caesar* that 'when beggars die, there are no comets seen; The heavens themselves blaze forth the death of princes'.[1] Comets have been held to be responsible for such different phenomena as the paranoia of the Emperor Nero, the collapse of the Aztec empire, religious revivals and mass suicides. Halley's Comet was visible from April to June,

1066 AD just before the Norman Invasion of England and the apparition is recorded on the Bayeux Tapestry. The comet was widely taken as an omen, but it was not clear what disaster it portended until the defeat of Harold at the Battle of Hastings in October of that year. Then it was apparent that William The Conqueror's claim to the English throne had heavenly approval.

The great French student of the solar system, Laplace, whom we frequently encounter in these pages, commented in 1796 that

> the appearance of comets, followed by these long trains of
> light, has for a long period terrified mankind, always agitated
> by extraordinary events of which the causes are unknown. The
> light of science has dissipated the vain terrors that comets,
> eclipses, and many other phenomena inspired in the ages of
> ignorance.[2]

Many comets are observed each year from the Earth. Once a comet is dragged into a close approach to the Sun, heating drives off the water, other gases and dust, so forming the spectacular tails. Comets thus survive for only a short time, measured in tens to hundreds of thousands of years. This is like 'a watch in the night'[3] compared with the age of the solar system, which has lasted a million times longer. So there has to be a regular supply of these short-lived objects from some distant and plentiful reservoir. Comets remind us of Icarus, who flew too close to the Sun. He came to grief when the wax that held his wings together melted. Comets last somewhat longer than that unfortunate early flier, but eventually die from the same cause, too close an approach to the Sun.

Comets are fragile enough; dirty snowballs as Fred Whipple (b. 1906), the American astronomer who was the first to appreciate their true nature, called them in 1950. Like snowballs, they fragment easily, particularly if they stray within the grasp of a giant planet. Comet Shoemaker–Levy provided a good example. It was captured into an orbit around Jupiter in 1929. For the next 65 years, it slowly spiralled in until July, 1994, when the comet broke up into about 25 pieces due to the gravitational pull of Jupiter. In another example, Comet Brooks came too close to Jupiter in 1889 and broke into at least nine pieces.

A disk of comets

One class of comets frequently reappears in our sky, typically at intervals between five to 20 years. These so-called short-period comets come from a disk of material just beyond Neptune and have been perturbed into planet-crossing orbits by the gravitational forces of the giant planets. Nearly 200 have been recorded, and their orbits lie close to the plane of the solar system. Their source is called the Edgeworth–Kuiper Belt after K. E. Edgeworth (1880–1972) and Gerard Kuiper (1905–1973), who first conceived of this notion. The

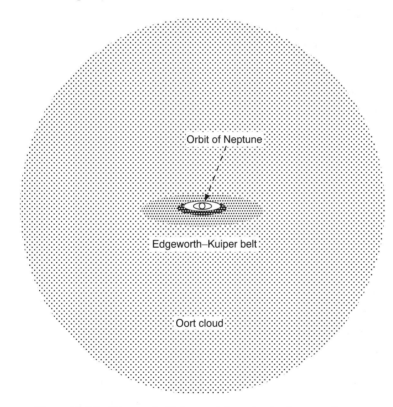

Figure 20. The Edgeworth–Kuiper and Oort clouds of comets that surround the solar system. Despite the immense region that the Oort Cloud occupies, extending out halfway to the nearest star, the total mass of the billions of comets is probably only a few times that of the Earth.

Belt is located out beyond Neptune, extending from about 30 to well beyond 1000 AU (see Figure 20).

Nearly all the short-period comets go around the Sun in the same sense as the planets. The belt contains perhaps 70000 comets larger than 100 kilometres in size, with perhaps several million bodies of the order of ten kilometres in diameter, that is, around the size of Mt Everest, but much less dense than that peak of Permian limestone. However, all the cometary material doesn't amount to much. It probably adds up to one quarter of an earth mass at most. Spacecraft that have now penetrated to these remote reaches do not detect any gravitational pull that would signal the presence of much material. Nor have they detected any large planets past Neptune, for that matter.

In the nineteenth century, asteroids were discovered in vast numbers between Mars and Jupiter. Now, as technology has improved, we are finding an increasing number of icy bodies at the inner edge of the Edgeworth–Kuiper Belt just beyond Neptune. The largest (1996TL66) is a smaller cousin of Pluto nearly 500 kilometres in diameter. It has a wildly eccentric orbit that extends from just beyond the orbit of Neptune out to 130 AU, much further away than Pluto.

A spherical cloud of comets

As well as those short-period comets that reappear every few years, there is a second major class of comets. They take long periods, often over 200 years, to reappear in our skies. Many appear only once on historical time scales, with their next appearance scheduled far in the future when they have completed an orbit that might extend out to thousands of AU. Unlike the short-period comets, they come from much further away and their orbits are often inclined at steep angles. Their source lies in a spherical cloud of comets, called the Oort Cloud, after Jan Oort (1900–1992), who predicted its existence in the 1950s. This cloud of comets surrounds the solar system like a halo and extends about a quarter of the way to the nearest star. The total number of comets in the Oort Cloud is uncertain but it may contain of the order of a trillion comets. Despite this enormous number, the amount of matter out there probably amounts to only a few times the mass of the Earth.

Comet Halley

Comet Halley is known to everyone. It is an example of a long-period comet, appearing about every 76 years, rather more frequently than most. This period is close enough to a human lifespan so that some people have seen it twice. Edmond Halley (1656–1742) saw the comet in 1682 and realised that the comets that had appeared in 1456, 1531 and 1607 were reappearances of the same visitor. He predicted that it would return in 1758. He did not live long enough to see his prediction fulfilled. However, he hoped that, 'if the comet should return in 1758, candid posterity will not refuse to acknowledge that this was first discovered by an Englishman'.[4] His wish has certainly been fulfilled.

Halley has been perturbed out of the Oort Cloud into an earth-crossing orbit that is in the opposite sense to those of the planets. The chunk of ice and dust that we call the nucleus was seen by spacecraft during the 1986–87 encounter (see Figure 21). It turned out to be a

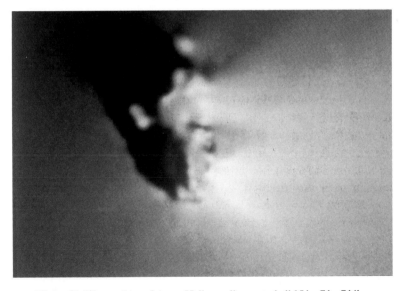

Figure 21. The nucleus of comet Halley, a dirty snowball 15 by 7 by 7 kilometres, spewing out gas and dust as it was heated when it neared the Sun in 1986.

body shaped like a potato or a peanut, according to taste. It is 15 kilometres long and seven across. This sighting confirmed the dirty snowball model of Fred Whipple. Halley, a veteran comet, was spinning slowly, rotating once a week. Most of the surface is coated with dark tar-like material, a residue of organic compounds. The spacecraft observed that gas was spurting out from several jets on the nucleus. About 20 tons of gas (mostly water) and several tons of dust were lost every second. These rates vary widely. Thus, Comet Hale–Bopp lost 200 times as much dust and 20 times as much water during its close approach to the Sun in 1997. Halley loses about one thousandth of its mass at each close approach to the Sun. It is calculated to have spent about 23 000 years in the inner solar system. During that time it has gone round the Sun perhaps 300 times. It should survive for another few hundred orbits. Comet Halley will thus last about a hundred thousand years since it was dragged out of its home in the Oort Cloud. This time is short on cosmic scales but long compared with human lifetimes. By the end of its life, it will either have evaporated completely, or become a small dark body that has lost most of its ice. A number of these dead comets are suspected to be present among the asteroids that approach, and may collide with, the Earth.

Comets from outside the solar system?

It has sometimes been suggested that comets come from outside the solar system. As many comets finish up being tossed out of our planetary system completely, if other planetary systems exist similar to our own, then comets might be thrown in our direction as their giant planets evict planetesimals. Interstellar space might thus be heavily populated with comets.

Such comets coming from interstellar space would be revealed by a hyperbolic orbit. None have been definitely confirmed so far, but the question is open. Their apparent absence at present might mean that comets are rare in interstellar space, or perhaps it is an observational defect. A different planetary system, perhaps consisting of small planets only, might lack the ability to throw out comets.

Although comets coming from outside our system have yet to be confirmed, some dust grains do arrive from these remote reaches of space. Although 99 per cent of dust grains that are observed to burn up in our atmosphere as shooting stars come from within the solar system, somewhat less than one per cent have velocities so high (greater than 100 kilometres per second) that they must come from outside our solar system. These dust grains are around 20 microns in size and seem to have come from nearby young stars.

Are comets primitive samples of the solar nebula

It was widely anticipated that comets would prove to be primitive unaltered samples of the original disk of dust and gas from which the solar system formed. This was based on the wishful thinking that leads us to consider that unknown regions are both uniform and simple. The ancient Greeks had similar ideas, but in mediaeval times, *terra incognita* was populated with monsters. The modern scientific viewpoint has rejected such fantasies. It has tended to the opposite extreme and considers these regions that we cannot easily observe, such as the deep interior of the Earth, or the primitive solar nebula, as uniform.

There has been a general feeling that comets are likely to provide the best sample of the original nebula. Coming from the outer reaches of the nebula, surely nothing much could have happened to them in that deep-frozen dim solitude. Thus, they might represent nearly unaltered samples of the fragment of the molecular cloud that was the original source of the gas and dust disk. A sample from inside the nucleus of a comet might thus be the long-sought primitive sample.

Comet Halley was expected to provide such answers. However, it appears unlikely that it is the long-awaited pristine sample of the primitive nebula. The refractory elements in the dust show large differences from the abundances of the chemical elements in the Sun. The iron and magnesium abundances in Halley are the lowest measured on any body in the solar system. These data from Comet Halley indicate that the composition is far removed from those either for the primitive meteorites or for the Sun. The nucleus itself is heterogeneous on a local scale. Instead of providing a primitive 'Holy Grail' sample of the nebula, Comet Halley is thus providing new insights into the complexities of chemistry in the nebula.

Our new discoveries of many bodies in the Edgeworth–Kuiper

Belt of comets had shed some light on this problem. Collisions are likely to be frequent in the belt and assist in sending bits earthwards. Thus, most of the comets that we see, typically a few kilometres in size, are likely to be broken off bits from larger bodies. Heating during these collisions may produce changes in composition. Finally, as the comet passes near the Sun, additional heating may modify the chemistry. Thus, there are probably as many differences among comets as in most other natural phenomena.

This search for the pristine sample, like those for other El Dorados, appears to have been based on false assumptions. However, such searches, although they have usually been conducted for the wrong reasons, have often produced unexpected benefits and fruitful discoveries. The search indicates that our previous motivation about the primitive nature of comets was misjudged. This is a common human condition. We are likely to obtain unanticipated benefits from future cometary missions, which gives a new impetus to sample these bodies. Like the seekers after El Dorado, we have stumbled across something else.

How do comets form?

So what was the origin of these these great clouds of comets? As dirty snowballs, they are clearly a mixture of ice and dust that must have formed far from the Sun where water ice was stable. However, there was never enough material in the original disk of dust and gas for it to extend as far as the present site of the Oort Cloud. At a little beyond Neptune, the disk became very thin, effectively running out of material. Neptune and Uranus, having grown to become ice giants, began to behave in typical giant fashion, tossing out of their way all the icy bodies they couldn't get hold of. The gas giants Saturn and Jupiter assisted in this gigantic ball game. They behaved like boisterous giants, tossing many icy bodies right out of the solar system. In this manner, the outer reaches of the solar system were populated with a great cloud of comets.

Such clouds of comets need not be a fixed or inevitable feature of planetary systems. They depend on the formation of giant planets. Comets as we observe them are characteristic of mature planetary systems very like our own. They are not likely to appear in a growing planetary system, since the following steps have to occur. Firstly, condensation of icy bodies has to occur in the disk. The next requirement is the formation of giant planets. These act as tossers out, scattering the comets out to a distant reservoir. As comets approach the Sun, heating produces the highly visible coma and tails. Only at that stage will comets become clearly visible objects. Accordingly, comets are not to be expected to be seen until the planetary system has reached maturity. Many chance events are needed to produce them.

Closely related to the comets are Pluto, Triton and the Centaurs, which I discuss next. All were probably former members of the Edgeworth–Kuiper Belt of comets.

The ice dwarfs and the centaurs

The ninth planet?

Tiny Pluto is commonly referred to as the ninth planet. The mass of Pluto, even when Charon is included, is very small. It amounts to less than one fifth of the mass of the Moon, $1/2000$ of the mass of the Earth or $1/64000$ of the mass of Jupiter. Pluto has a highly inclined and eccentric orbit. Sometimes it is inside the orbit of Neptune. It will move out past Neptune in March, 1999. In the frozen twilight, nitrogen ice lies on the surface of Pluto. As Pluto get closer to the Sun, it warms up a little. The nitrogen evaporates and forms an atmosphere. As Pluto retreats from the Sun, the atmosphere freezes out again.

Although the orbits of Pluto and Neptune cross one another, close encounters are avoided. Pluto orbits around the Sun exactly three times for each two orbits of Neptune. Pluto was probably nudged into this safe orbit by a collision with another body, and this stabilised the orbit. Otherwise, Pluto would long ago have been swept up by or captured by Neptune, or sent sunwards.

It is apparent that Pluto is not a planet, although no doubt it will long continue to be referred to as the ninth planet for a combination of traditional and sentimental reasons. We will just have to put up with a solar system that has only eight planets, despite much hope for ten (a tidy number) or more, a wish that goes back to Kant. The ancients were content with five, as well as the Earth. Pluto is a cousin of Triton, but, like most relatives, is not identical. It is both smaller, a little darker and denser, probably the result of a different history. Pluto has a higher content of rock than Ganymede, Callisto or Titan, so calling Pluto an ice dwarf is a bit of a misnomer. It's yet another example of the resistance of objects in the solar system to being put into neat pigeonholes. I will explain shortly how it may have lost some of its ice.

Finally, at least 40 bodies, of which 1992 QB1 was the first to be discovered, have been found in orbits beyond Neptune. They are the long-predicted inhabitants of inner edge of the Edgeworth–Kuiper Belt of comets. They show some similarities to Pluto, strengthening the family resemblance.

The strange case of Pluto and Charon

Pluto has a large satellite, Charon, discovered only in 1978. The orbit of Charon about Pluto is bizarre, being inclined like the satellites of Uranus, at about 90 degrees to the plane of the solar system. Charon and Pluto remain fixed relative to each other in their dark skies, looking at each other's face in an eternal gaze. Charon is one seventh of the mass of Pluto, making it the biggest offspring, relative to its parent, in the solar system. It far exceeds that of the nearest rival, the Moon, that is only 1/81 of the mass of the Earth. The masses of the satellites of the giant planets are trivial compared with their mighty parents. Charon is much less dense than Pluto and contains a large amount of ice, whereas Pluto is more rocky.

Pluto and Charon, in addition to their other strange properties, are spinning even more rapidly than the Earth–Moon pair. Probably another body slammed into Pluto, and Charon was born from the collision. This impact could have stripped the ice off Pluto that finished up in Charon.

The origin of Pluto

Many theories have been proposed for the origin of Pluto. Because of its density, it was sometimes thought to have formed in the inner solar system. The difficulties in transporting such an object from the inner to the outer reaches of the solar system are considerable. Obstacles in the way include the zoned structure seen in the asteroid belt and the presence of Jupiter and the other giant planets. Since there was no water ice this side of Jupiter, it seems an unlikely place to try to make the tiny cold dwarf in the first place.

Another popular notion was that Pluto was a satellite that escaped from Neptune. This idea has been mostly discredited because of the immense difficulties involved in separating the little fellow from the clutch of the ice giant.

The most likely explanation is that Pluto represents a large icy body that formed in the outer solar system. Thus, it is really a big comet, a cousin of Triton and Chiron. However, it has too much rock, or too little ice, depending on how you view it, to be a typical denizen of the region. Somehow it got rid of a lot of the ice, probably during the collision that formed Charon.

Such a history must make us cautious about using the composition of Pluto to tell us about the composition of the primitive disk of dust, ice and gas. Probably neither Pluto nor Triton represent a fair sample of the early solar system. Like ancient crones, too much has happened to them since early childhood and neither is truly primitive. Pluto and Charon are so far from the Sun that at that distance it is only 1/1000 times as bright as we see on Earth. Nevertheless, this pair have managed to have just as an eventful history as the more sunward members. So the solar system does not become simpler with increasing distance from the Sun.

The capture of Triton

Triton is the large satellite of Neptune. Henry Cooper, writing about the *Voyager* encounter in the *New Yorker* magazine, commented that instead of being the expected featureless snowball, 'Triton, with its tectonic features, its crisscross of channels, its mushrooms, its wind streaks, its haze, its evidence of condensing volatiles, was beginning to look like Europa, Enceladus, Mars, and Io rolled into one'.[5] It is considerably smaller than the Moon but significantly larger than Pluto. It is orbiting backwards around Neptune in a nearly circular but highly inclined orbit, the only large satellite with such an orbit.

The thin atmosphere of Triton is mainly composed of nitrogen, similar to that of Titan, the giant satellite of Saturn. The temperature on the surface is only 38 degrees above absolute zero. The surface slowly changes with the seasons. At present the south pole of Triton faces the Sun. It is covered with blue nitrogen ice that is slowly warming and evaporating as the Sun heats it. There are geyser-like plumes of nitrogen in active eruption that are about eight kilometres high. When they reach that altitude, the plume trails horizontally, like smoke from a factory chimney encountering a temperature inversion. There is about one kilometre of relief on the surface. Under the coating of nitrogen ice, it is probably made of water ice. Parts of the surface are dimpled like a gigantic melon hanging in the sky except that the dimples are nearly 50 kilometres across. Much of the surface resembles volcanic terrains on the Earth, except that the 'lava flows' are mixtures of ammonia and water ice.

Triton was almost certainly captured into a backward orbit by Neptune. During this traumatic episode, it probably collided with a satellite of Neptune and demolished any previous regular satellite system around the planet. Triton seems to have undergone some form of plastic surgery as it shows a comparatively youthful face. Probably, Triton was melted during its capture and remained hot for five hundred million years after that event. By the time that the surface froze again and became strong enough to preserve the record of impacts, the period of heavy bombardment in the solar system was over. So unlike on the Moon, Mercury or Callisto, there is no old heavily cratered surface. The largest crater seen so far is only 27 kilometres in diameter.

Both Triton and Pluto have been altered from their original state by a violent history. Pluto was hit by another body, while Triton melted following its capture by Neptune. It is not surprising that these cousins now look a little different, like family members who have gone off in different directions. There must have been many other such bodies, perhaps ten thousand such icy dwarfs, now swept up into the giant planets, or tossed out to the Edgeworth–Kuiper or Oort clouds or out of the solar system entirely. Thus, there is a close connection between comets, Pluto, Triton, Edgeworth–Kuiper Belt comets and the centaurs, which I now investigate.

A herd of centaurs

The centaurs were beasts in mythology that were half man and half horse, just about as strange as their namesakes in the solar system. The first sighting of these creatures was in 1977, when Chiron was found, wandering apparently alone in the immense void of ten AU, between Saturn and Uranus. This was a surprise, since it used to be thought that the space between the giant planets was clean, an idea going back to Newton. Chiron, about 175 kilometres in diameter, is one of the most isolated bodies in the solar system. It is a dark grey-black chunk, probably coated, like Comet Halley, with dark tar-like material. Occasionally, it is seen spewing out some gas. This makes it a large comet by definition. Chiron's closest approach to the Sun is inside the orbit of Saturn, while its orbit extends out almost to Uranus. Chiron is very similar in size and colour to Phoebe, one of the captured satellites of Saturn. Probably some of the captured satellites of the giant planets were once free-roaming centaurs. Sooner or later, within a million years or so, Chiron will collide with Saturn or perhaps Uranus. It has enough mass to duplicate the beautiful ring system of Saturn, if it were captured into the right orbit. Hopefully, it will not penetrate into the inner solar system on its chaotic path.

The next member of this wandering herd, found 15 years later, was Pholus. It turned out to be another icy body, in an orbit that extends from that of Saturn to out beyond Neptune. Like Pluto, its orbit is highly inclined to the plane of the ecliptic. It is a little larger than Chiron and is probably an extinct comet with an armoured surface of organic material. It last came close to Saturn in 763 BC, ten years before the founding of Rome.

A few other similar bodies, such as the well-named Damocles, have been seen, but they are hard to spot. The herd of centaurs may number a few hundred. Probably they are wanderers that were shaken loose from the Edgeworth–Kuiper Cloud. They provide an interesting contrast to the asteroids, which are typical rocky inhabitants of the inner solar system. Such rocky bodies went to make up the Earth, Venus, Mars and Mercury. In contrast, Triton, Pluto and the rest are icy bodies from far away. They must resemble the building blocks of the cores of the giant planets.

Like any herd of wild horses, they have an unstable existence. They will finish up becoming short-period comets, or being captured or colliding with one of the giant planets. If the angle of the encounter is just right, the bits and pieces of rubble can end up forming rings around the planet, a topic to which I now turn.

Planetary rings

An early puzzle

The beautiful and exotic rings of Saturn were for a long time the only known examples of a system of rings around a planet. Laplace commented in 1796 that 'Saturn presents a phenomenon unique in the system of the universe'.[6] Certainly, the first view of Saturn through a small telescope reveals an improbable sight. It is almost as bizarre as if one saw a Chinese lantern swimming in space. The rings have excited interest and wonder ever since their discovery by Galileo. He thought that there were two large satellites, one on each side of the planet. In another view, the rings were thought to be lobes, attached like ears to the planet. Then Christian Huygens (1629–1695) discovered that they formed a large ring which was completely separated from the planet.

What was it? We now are so used to disks in nature, in the form of hurricanes and spiral galaxies that it is difficult to imagine the intellectual effort involved from changing from models with spheres to those with disks. The ancient Greeks had thought that the proper shape for the heavenly bodies was a sphere. The spherical shape of Sun and Moon are obvious to all. After all, disks are uncommon on human scales. They are mostly only clear when viewed from a distance. The disk form of hurricanes is not apparent to ground-based watchers. Sophisticated equipment is needed in order to detect the true spiral form of hurricanes or spiral galaxies. Huygens thought that Saturn's ring was thick and solid. Even as eminent a scientist as Laplace, followed the opinion of Huygens and thought that it was a solid disk, looking like a sort of celestial phonograph record. Earlier, Kant correctly supposed the rings to be composed of many small particles, each one circling the planet in accord with the rigid laws of dynamics. Later workers confirmed Kant's notion.

Primitive survivors?

The rings have usually been thought to have some fundamental significance for the origin of the solar system. Saturn's rings are so extensive that they have often been thought to be primitive, a left-over disk from the formation of the planet, so giving clues to the breakup of the original disk of gas and dust into planets. In the nineteenth century one worker commented that 'Saturn's rings were left unfinished to show us how the world was made'.[7] The idea still lingers around. The apparent absence of similar rings around the other giant planets was something of a mystery, but perhaps could be explained.

Things became more complicated when, within a few years, rings were found around the other major planets. On March 10, 1977, the rings around Uranus were discovered, followed two years later, on March 4, 1979, by the ring system around Jupiter. In the succeeding two years the extraordinary detail of the saturnian ring system was revealed. *Voyager 2* made a close examination of the nine rings around Uranus in January, 1986. It viewed in August, 1989, the clumpy rings around Neptune, which had been detected shortly before from Earth.

Compared with the rings of Saturn, the newly discovered rings are mostly darker and much less massive. It was these factors that hindered their discovery until very recently. However, the puzzle of their origin became worse, for the rings are all different. They do, however, share one common property. They all lie close in to the planets where gravitational forces are sufficient to tear apart any fragile body trapped within the embrace of the giant.

The Lord of the Rings

The splendid rings of Saturn have now been resolved in spacecraft photos into many thousands of rings of particles circling in stately orbit around Saturn (see Figure 22). Each particle takes about a day to orbit the planet. Although the ring system is of vast extent (the diameter is close to the Earth–Moon distance) the rings are extraordinarily

Figure 22. The rings of Saturn, viewed by the *Voyager* spacecraft after it had passed Saturn. The Sun is to the right and part of the rings is in the shadow of the planet. (NASA *Voyager* P-23254).

thin. The true thickness is certainly less than 50 metres. Perhaps they are only a few metres thick. A fairly accurate scale model would be a film of plastic food-wrap spread smoothly over a football field. The average particle size in the rings is only a few metres, although some house and mountain-sized objects are present as well among the rubble. The ring particles reflect sunlight very strongly, since they are coated with water-ice. Probably they are mostly ice, rather than frosty rocks. Fine dust is also present, particularly in the outer rings, and probably comprises all of the ethereal fairy-like E ring (the major divisions of the rings are conventionally labelled as A, B, C, etc. in the order of their recognition by early observers). The E 'ring' is more like a cloud than a disk. It spreads vertically for a few thousand kilometres, in contrast to the extremely thin main rings. It seems to be being formed by a spray of fine dust that results either from volcanic activity on the nearby satellite, Enceladus, or from the collisions of small ring particles with the icy satellite.

A striking observation is the sharpness of the edges of the individual rings of Saturn. The B ring diminishes from being very dense, packed with boulders, to nothing over a distance of a little more than one kilometre. The edges of other rings are less than 100 metres wide. Possibly the ring edges are kept sharp by small shepherding satellites. These are mostly too small to have been discovered at present. The analogy of shepherds (small satellites) controlling flocks of sheep (ring particles) is frequently used. However, sheepdogs would be a better term, since they are much more efficient than shepherds. This is a piece of wisdom inherited from my youth on a New Zealand farm.

Thinner and darker versions

Jupiter has a single very thin ring. It is made of particles only a micron or so in diameter. Such an ethereal ring reminds one of Tinker Bell, the nearly invisible fairy friend of Peter Pan and Wendy. The ring needs a continuous supply of particles, as such tiny particles spiral into Jupiter fairly quickly. The source of these is probably one or more tiny moons that have not yet been found.

Uranus has nine very dark thin rings. They are not as black as commonly painted, but are dark grey and very opaque, unlike the rings of Jupiter and Saturn. The ring particles are small, again about one micron in size. The individual rings are between one and twelve kilometres wide. They have sharp edges, which seem to be sculptured by shepherding satellites. In between them are some almost invisible broad dusty lanes. The rings, like those of Saturn, are also very thin, being only a few metres thick. However, the rings of Saturn contain a thousand times more material.

Five rings have been identified around Neptune (see Figure 23). Two of these are named for Galle who found Neptune, and Le Verrier, who told him where to look. The outer ring is named for Adams, who had also calculated where to find the planet, but who failed to persuade the astronomers in England to look quickly enough. So the three people most closely associated with the discovery of the planet Neptune finally have their reward. The Adams ring is famous not because it is uniform, but because it has three thicker sections. One might have expected that material should spread around the ring in a period of a few years rather than remain clumped together; the clumping is probably due to the shepherding influence of a nearby satellite, Galatea. The amount of material in the neptunian ring

Figure 23. Two rings of the Neptune, seen from *Voyager 2* at one million kilometres from Neptune. The inner or Le Verrier ring is 53 000 kilometres distant from Neptune and the outer, or Adams ring is at 63 000 kilometres. The three clumpy arcs in the Adams ring are clearly shown. The rings are bright in this view, due to scattering of light from the microscopic dust from which they are formed (NASA JPL P-34712).

system is very small. It comprises only about one per cent of that of the rings of Uranus, or 100 000 times less than the magnificent rings of Saturn.

What about the inner planets? Searches for rings around Mars have revealed no trace of a ring inside the orbit of Phobos, where one might be expected. Neither are there any signs of rings around Mercury, Venus or the Earth

The origin of the rings

The discovery of rings around Uranus, Jupiter and Neptune, as well as the presence of the famous rings of Saturn, naturally raises the question as to whether the formation of rings is inevitable during planetary formation. The contrast could hardly be greater between the different ring systems. It is not related to the size or position of the planet. Jupiter has a thin ring, composed of dust, Saturn has its magnificent broad bright rings, Uranus has its dark narrow rings, with some dusty lanes, while Neptune has clumpy rings. Workers have commented that as the new data appeared, 'the observations once again seem to fall just short of revealing the essential nature of rings'.[8]

Clearly no uniform ring-making process has been in operation. There is so much variety among the rings, as with everything else in the solar system, that it is difficult to extract any general rules. What are we to make of all this diversity in mass, structure and composition? There are a few clues. Firstly, there is not much material in the rings. All the material in the rings and shepherding satellites of Saturn could be contained within an icy satellite about 200 kilometres in radius. A small dark body would contain the material in the rings of Uranus. The total mass of the Jupiter ring or of those of Neptune is trivial.

As we saw above, the rings all lie close in to the parent planet, within about three radii of the planet. This is the famous limit worked out by the French mathematician, Edouard Roche (1820–1883). Any fragile body that comes this close within the embrace of the giant planet will be torn apart, like a giant seizing a fairy in a sinister tale for children. The Shoemaker–Levy comet, which was broken into about 25 pieces by Jupiter, is a dramatic example. If it had been on a different orbit, it might have formed a dusty ring, rather than having the bits and pieces collide with the planet in such a dramatic fashion. Theory also predicts that the rings should not be stable over long periods. The particles should be swept up by the parent planet over periods of the order of a few hundred million years. This is much shorter than the four and a half billion years that the solar system has been around. If the calculations are correct, the rings have a relatively short life.

So perhaps the rings formed from broken-up pieces of satellites that strayed within the gravitational grip of the planet. There are lots of satellites of the right size. The basic problem with this model is that the formation of the satellites belongs to a much earlier period, right back at the beginning of the solar system. They have long since settled into stable orbits. Like well-behaved people keeping clear of danger, they are no longer likely to be caught straying near the planet.

The more likely picture is that the rings are produced by the breaking up of captured comets, such as Chiron. The tiny particles become the rings, and the bigger bits make tiny moons. These act as shepherds or sheepdogs, according to taste. They also supply dust as they are ground up by collisions among themselves, like boulders in a stream bed.

The beautiful rings of Saturn may thus be disrupted fragments of icy comets. The more ethereal Tinker Bell type rings may be real transients. The different masses and colours of the rings are a natural outcome of the breakup of bodies of different compositions, some icy and bright, some dark and rocky.

A passing scene

If the breakup of captured comets is the correct explanation, then the spectacular rings are late-comers to the solar system. They are accidental features and have no fundamental significance for the origin of the solar system, except to reinforce the general model that the planets formed from smaller bodies. The clumpy nature of the Adams ring of Neptune probably demonstrates the temporary nature of ring systems on solar system time scales. Thus, like so much else in the solar system, the planetary rings are due to chance events, and form when a small cometary body wanders within the gravitational embrace of a giant planet. The absence of rings around the terrestrial planets is due to their small size. Comets and asteroids mostly collide directly with these planets and explode to form craters. Any broken-up debris is fairly quickly swept up by the planet rather than forming an extended ring.

How often will such events happen? Estimates vary but perhaps a hundred close encounters of large comets with giant planets could occur over the age of the solar system. Most would leave no material in orbit, since that will require an impact at just the right angle and velocity.

Following the breakup, it should only take between ten thousand and a hundred thousand years for the material to spread out in a disk. Possibly only one ring-making event will occur for each planet over the whole age of the solar system. This model seems consistent with the observations of very distinct sets of rings around each of the major planets. It also means that since the life span of the rings is a few hundred million years, they are unlikely to be renewed. Thus, the rings seem to be relatively short-lived. We must judge ourselves fortunate to be present in the solar system at the same time as the splendid rings of Saturn. Perhaps we should appreciate them even more, since they may well be, like so much else around us, unique in the universe.

The asteroids

The vermin of the sky

The asteroids have been a great source of irritation to astronomers, who are interested in more distant objects. This swarm of bits and pieces gets in the way, leading to the insulting title of 'the vermin of the sky'. However, they have a tale to tell to those interested in the origin of the solar system. The major questions are why there is so little material and no planet at the Titius–Bode rule position between Mars and Jupiter. This gap in the orderly sequence of planets had been noted ever since the time of Kepler. When Johann Daniel Titius (1729–1796) came up with his famous rule in 1766, it was clear that a planet was missing between Mars and Jupiter. After Herschel discovered Uranus in 1781 at 19.2 AU, close to the position at 19.6 AU predicted by Titius and Bode, astronomers began to take the search seriously. Curiously enough, Giuseppe Piazzi (1746–1801), who was merely checking a star catalogue on New Year's Day, 1801, discovered Ceres at 2.77 AU, close to the predicted spot at 2.8 AU, but lost track of it soon after. He did not live long to enjoy his fame, dying the same year, perhaps overcome by his discovery.

Ceres, number one in the asteroid catalogue, is the largest of its kind, containing about one third of the mass of the entire asteroid belt. Its diameter is 933 kilometres. The next one, Pallas, was discovered in 1802 by Wilhelm Olbers (1758–1840), whom we last met discussing why the sky is dark at night. He also managed to relocate Ceres. Juno and Vesta were found soon after. Olbers thought that these tiny bodies were fragments of an exploded planet, an idea that still lingers around. Laplace, to his credit, considered that they were bits of an unfinished planet, more or less in line with modern thinking.

A great multitude

It was not until 1845 that another asteroid, Astraea, was discovered. The floodgates then opened. Now over 10 000 have been identified. There is an uncounted multitude of smaller bodies less than a few kilometres across. The orbits of about 6 000 asteroids are well enough known for them to be assigned numbers. Although in popular mythology the asteroid belt is portrayed as a swarm of boulders that would be a hazard to intrepid space travellers, they are mostly separated by millions of kilometres and don't add up to much. The total asteroid mass amounts to only five per cent of that of the Moon. They have interesting and often whimsical names that I make no apology for mentioning. Typical asteroids are Ida and Gaspra, respectively numbers 243 and 951 in the catalogue. Both were photographed by the *Galileo* spacecraft. Asteroids are mostly spinning rapidly due to collisions. They have rotation rates of a few hours. The average period is about eight hours. Ida, seen close up by the *Galileo* spacecraft (see Figure 24), has a tiny moon, Dactyl, about a kilometre and a half in diameter.

There are three main groups of asteroids. Most lie in the Main Belt spread out for over an AU around the Titius–Bode position at 2.8 AU. Closer to Mars are the Hungarias, named after the principal member of the group. Beyond the main belt are the Cybeles, named for the most prominent member, at 3.3–3.5 AU. Farther out are the Hildas. Another group, the Trojans, are scattered around the same orbit as Jupiter. Like camp followers keeping well clear of danger, they occupy two stable positions, named after the great French mathematician, Comte Joseph Louis Lagrange (1736–1813). One group precede Jupiter, like outriders warning of the approach of the giant. A second crowd of Trojans trail the planet as it makes its stately eleven year orbit around the Sun.

Of more immediate concern to us are the Near-Earth Asteroids, whose orbits cross those of the inner planets. They are divided into the Apollos, Atens and Amors (see Figure 25). The Apollos, despite their lovely names, have the capacity to do us deadly damage, as they

Figure 24. Ida, a typical asteroid, 56 kilometres long. It has had a long history of collisions shown by its pock-marked face (NASA).

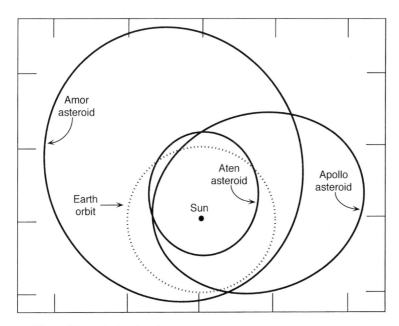

Figure 25. Typical orbits of the Apollo, Aten and Amor Near-Earth asteroids.

are in earth-crossing orbits. There are estimated to be 1000 of them that are more than one kilometre in size, with the potential for widespread devastation if they hit the Earth. Only a couple of Apollos, Sisyphus and Phaeton, approach the size of the body that extinguished the dinosaurs 65 million years ago; that was a spectacular case of bad luck for them, although not for us. The Atens have orbits that lie within that of the Earth, but cross our orbit at their maximum distance from the Sun. The Amors have orbits that cross that of Mars, and approach, but do not cross that of the Earth. The largest one, Ganymed, not to be confused with Ganymede, the big satellite of Jupiter, is over 38 kilometres in diameter. Another member of the Amors, called Eros, 22 kilometres in diameter, is on the list to be visited by a space probe.

The source of meteorites

The asteroids are rocky and metallic debris. We know this because they are the source of most meteorites, excepting those few that arrive from Mars or the Moon. There are gaps in the belt of asteroids. These are called Kirkwood Gaps after the Indiana University astronomer, Daniel Kirkwood (1814–1895), who discovered them last century. These occur at simple ratios of the orbits of the asteroids with Jupiter. At these locations, the gravitational force of this giant focusses to toss some of them in our direction so providing us with meteorites. Thus, these informative visitors are sent to us courtesy of Jupiter.

The most valuable contribution that the asteroids make to our understanding is to provide us with samples of the early solar system. We can discover both their composition and their age in terrestrial laboratories. However, there are a large number of asteroid compositions not recognised among the present collection of meteorites. We keep finding new varieties in Antarctica and in deserts. So we should be cautious in placing too much reliance on the present meteorite collection as complete. However, although our samples are surely biased, without these samples of the asteroids we would be hard pressed to understand the early history of the solar system.

Classes of asteroids

There are many distinct types of meteorites, so it is not surprising that the asteroids also vary widely in composition. About a dozen different varieties have been recognised through the rather difficult art of studying the light reflected from them. Those on the sunward side of the main belt are mostly mixtures of metallic iron and rocky minerals. Others appear to be the iron cores of broken-up bodies. Some of these are kilometre-size chunks of nickel–iron steel. They contain enough of the precious elements that would enable everyone to have platinum saucepans, if we could mine them. Mining asteroids in fact is probably within the reach of our technology.

Rarely, an asteroid has erupted lava on to its surface. Vesta is the famous example, and we have meteorites from that body that tell us of volcanic activity four and a half billion years ago, soon after the origin of the solar system. Further away from the Sun, more primitive compositions dominate. It is from this region that we have derived those meteorites that tell us about the composition of the solar nebula.

Like the rocks on Earth, which are familiar to every beginning student of geology, we can place the asteroids into three main types, 'igneous', 'metamorphic' and 'sedimentary'. The 'igneous' ones have been melted and are made of metal and rock. The minerals in the 'metamorphic' analogues have been altered by melting of water ice, while the so-called 'sedimentary' ones are compacted collections of dust and minerals, unaltered from the beginning of the solar system. We don't know what caused those nearer the Sun to be melted, but whatever it was, it was right back at the beginning and had something to do with the early Sun, or maybe some long-extinct radioactive element.

Families of asteroids

Although the exploded planet idea has long been discarded, except in some popular works, many asteroids are broken up bits of bigger ones, the results of collisions in the densely populated belt. These collections of fragments derived from the same parent body are called families and much work has gone to try to establish which asteroids are relatives. Enthusiastic early workers created more than a hundred families to which most of the known asteroids were assigned. Skeptics soon arose and cast doubt on the reality of most of these families. Recent studies suggest that only a few of the families are real and there appear to be less than about ten. Like the mythical canals on Mars, the erection of so many doubtful families is a cautionary tale for classifiers working at the limits of observation. Thus, many of the smaller 'families' are probably accidental groupings, while interlopers abound among the larger families. The family concept was a good idea that was pushed beyond reasonable limits. This is a common problem in the natural sciences.

A zoo or a wilderness area ?

One fundamental question is whether the asteroid belt is a trapped random collection of stray bodies, like animals in a zoo. Alternatively, is it a wilderness area, where the asteroids are preserved in their native habitat? If it is a true wilderness, then it can tell us something significant about the early solar system. In fact it looks as though the belt is truly ancient, and was there, more or less in its present state, from the earliest times. Unlike the rings around the giant planets, the asteroids and their belt are not of recent origin. Evidence for this comes from the ancient ages of most meteorites. The structure of the belt also appears to be very old. It does not seem to be a random mixture of the different kinds of meteorites. There is very little mixing among the different classes of meteorites. Few meteorites contain bits from other meteorite groups, so that mixing within the asteroid belt seems to have been relatively minor. The separate meteorite classes come from quite narrow zones, perhaps less than one tenth of an AU wide. So the asteroid belt is not a zoo full of trapped species, but has been there since the beginning of the solar system.

Origin of the asteroids

What are we to make of this great collection of debris? Why is it there, what is it due to and why isn't there a cousin of Mars sitting in the position predicted so long ago by Titius and Bode? Why did a large planet not grow in the asteroid belt? If there had ever been a large planet in that position, it would still be there. Why is there so

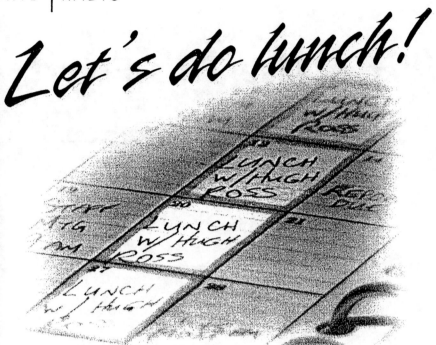

Please Use This Flyer!

We need help from people like YOU!

✔ People who know about Reasons To Believe

✔ People who can spread the word about
our exciting resources showing who
the Creator of the universe is

✔ Pass this announcement to a seeking friend

✔ Put it up on a bulletin board where you work

✔ Ask your church to use it as a bulletin insert

✔ Put it up on your church's bulletin board

✔ Feel free to make as many copies as you wish

*Thank you for partnering with us to show
Jesus Christ as the Master Designer!*

little material in the belt? This is the fundamental problem as there is no reason to suppose that the region of the solar nebula now occupied by the asteroid belt was initially so poor in material.

The swarm of tiny asteroids is there because of the early formation of giant Jupiter close at hand. This giant seized a lot of the bodies and tossed them far away. About half were thrown right out of the solar system. Perhaps a quarter finished up in the Sun, while a like fraction hit the Earth or Venus. Like a victorious army scattering a defeated foe, Jupiter then altered the orbits of the impoverished remainder. They were scattered so that they could not reform ranks and collect themselves into a planet. Like any fleeing crowd, many collisions took place between these survivors. We see the results on their broken and pock-marked surfaces. All these events took place within a few million years so that the belt in its present state dates back nearly to the origin of the solar system.

Since that time the asteroid belt has been a relatively quiet place. Asteroids stopped growing as the neighbourhood ran out of material. Although many collisions have occurred between them since then, this has not resulted in the building of even a tiny planet. The asteroid belt thus contains critical evidence relating to the origin of the solar system. Although it did not contain enough material to form its own planet, this fact sheds much light on the history of the solar system. Thus, the failure of the asteroid belt to produce a planet is of more value to this inquiry then if we had another version of Mars lying between that planet and Jupiter.

Other asteroid belts?

There is no sign of 'Trojan' type asteroids at the appropriate locations in the orbits of Saturn, Uranus and Neptune. If asteroids were widely scattered from the main belt by Jupiter, some might have been expected to be trapped by Saturn, Uranus and Neptune. None appear to be present, but perhaps they are too small to have been detected so far.

One asteroid, labelled 1990 MB, has been recognised, following along like a faithful dog behind Mars in its orbit. This is the first martian 'Trojan'. Another asteroid, number 3753, a tiny five kilometres across, has become trapped by the Earth into a kidney-shaped orbit that stretches from Mercury to Mars. This wandering path makes it a companion rather than a satellite of the Earth.

Other bodies wandering about in the outer solar system fall in the category of ice dwarfs and Centaurs that were discussed earlier. We have seen the fine-tuning required to make Jupiter. If it wasn't there, the asteroid belt would not exist. If Jupiters are rare in the universe, asteroid belts may also be uncommon. The inhabitants of such planetary systems will have to do without meteorites to inform them of their early history.

Mars

The red planet

Among all the planets, Mars has a unique fascination for the human imagination. This is because the surface conditions on this planet are more like the Earth than any other body in the solar system. It is a cold desert. The average temperature is 55 degrees below zero centigrade, or 218 K although it gets over 20 °C in the summer near the equator. We could imagine living there. Even the day length on Mars is close to that of the Earth. It would be a little more uncomfortable than living at the South Pole, with the snow blizzards replaced by long-lasting dust storms. Travellers would need oxygen, water, shelter and a good source of energy as well. Near the poles of Mars, if they were unwise enough to venture there, they could even run into flurries of dry ice.

The Romans identified Mars with the God of War, from its red colour, and it has exercised a strong hold on the human imagination. Thus, Mars has been a favourite location for science fiction, and the imaginary inhabitants of the red planet usually have had undesirable human characteristics. The invasion of the Earth by aggressive Martians in *The War of the Worlds* by H. G. Wells (1866–1946) was one of the first examples. Late in the nineteenth century, Giovanni Schiaparelli (1835–1910) observed regular channels crossing the surface of Mars. These he called canali, the Italian term for channels. Percival Lowell (1855–1916) took them to be canals, with vegetation along their banks. He finally mapped 437 of them criss-crossing the martian surface. The idea was that they had been constructed by a civilisation to bring water from the ice caps at the poles to the parched equatorial regions. The white polar caps do in fact have permanent water ice, which gets covered by a seasonal frost of carbon dioxide, our familiar 'dry ice'. The presence of the canals excited much interest, as they would constitute evidence for the existence of a technically advanced civilisation. The problem with observing them was that they were at the limit of the resolving power of the telescopes in use over a hundred years ago. Thus, they were seen by some observers but not by others. There was no question that the canals were produced by intelligent beings. The significant question is on which side of the telescope the intelligence was located, as the famous canals turn out to have been optical illusions. The martian canal story is another cautionary tale for scientists, illustrating the problems of interpreting data close to the limits of resolution.

Deserts of vast eternity

Mars is small, being only 11 per cent of the mass of the Earth. It is also significantly less dense. It is thus a poor relation both of the Earth and Venus. However, it makes up for being tiny by producing the largest landforms in the solar system. The largest mountains on the Earth appear as mere pimples compared with mighty Olympus Mons on Mars. This monster volcano rises to an elevation of 26 kilometres above the surrounding plains and spreads over a distance of 600 kilometres. The great valley of Mars, Valles Marineris, is 4000 kilometres long. It is so wide that the far rim would be beyond the horizon to martian travellers standing on the near side. The Grand Canyon of the Colorado River could be dropped out of sight into it.

Our travellers would have to protect themselves from the dust storms. Unlike those in deserts on Earth, they can last for months. This seems to be due to a combination of the very fine size of the dust grains, typically a few microns, and the low martian gravity. The large dust storms are responsible for the changes in the surface that are visible in Earth telescopes. These dark and light patterns of the surface were first thought to be due to seasonal changes in vegetation. The dust is typically about a metre thick on the surface. This is not deep enough to bury the rocks lying about, which have been mostly tossed there from meteorite impact craters. The dust arises mainly from the great basins in the south and is moved mostly from the southern to the northern hemisphere – the region sampled by the *Viking* landers. As the dust settles back onto the surface, it forms spectacular dark bands within the seasonal layers of ice at the poles.

Mars has naturally attracted attention as one of the few possible abodes for life elsewhere in the solar system. Two *Viking* Landers went in search. Nothing was found. Most significantly, there were no organic compounds, and a strongly oxidising surface that would be as effective as household bleach in destroying carbon-based life. Meanwhile, enormous interest has been generated by the reports of the possible existence of primitive bacteria on Mars. This evidence is contained in meteorites blasted to Earth from well beneath the surface. This entertaining topic will be discussed later.

Mars provides an endless list of questions. Why is it so small? The two landers, even though 4000 kilometres apart, found a similar surface of tumbled boulders of basaltic lavas. The volcanoes, although huge by our standards, have a similar shape to the Hawaiian lava dome of Mauna Loa. On the Earth, the mobile crust moves over the sources of volcanoes deep in the mantle. Thus, the volcanic Hawaiian Islands are spread out over a large stretch of the Pacific Ocean, a result of the Pacific Plate moving northwestwards over the hot spot producing the lavas at depth. On Mars, the crust is stationary, so that enormous volcanoes grow in one place, aided in reaching their great height by the low gravity of Mars.

A divided planet

There is a major division between the northern and southern hemispheres. The crust in the northern hemisphere, lying under a pink sky that is coloured by the red dust, consists mainly of monotonous lava plains.

The crust in the south lies between one and three kilometres higher than the rolling basaltic plains to the north. This southern crust is older and heavily cratered. It is a relic from the time of an early great bombardment over four billion years ago, which also heavily cratered the Moon and Mercury. Erosion has removed the evidence of this traumatic episode from the Earth, while Venus has buried the evidence under plains of lava.

The cause of the contrast between the northern and southern hemispheres on Mars has been much debated. Opinion veers between an internal source, and an enormous early impact, which dug a crater covering two fifths of the surface of the planet. Other workers call for several rather smaller impacts, for which there is some evidence across the northern plains. These kinds of differences occur elsewhere in the solar system. One is reminded of the difference in crustal thickness between the near and far sides of the Moon. However, it is hazardous to extrapolate from one planet to another.

A crust of lava

Although both *Viking* landers settled in the northern hemisphere, the fine dust analysed by them comes from all over the planet, blown by the dust storms. So the dust provides us with an average sample of the surface of Mars, just as wind-blown dust on Earth produces a tell-tale finger-print of our continents even in the muds on the deep floors of the oceans. From the analyses of the soils by *Viking* or *Pathfinder*, there does not seem to be anything on Mars that resembles the continents on the Earth.

The useful small rover on the *Pathfinder* Mission found a rock 'Barnacle Bill', which was labelled as 'andesite' and excited much interest. Although it was a bit richer in silica than either the *Viking* or *Pathfinder* soil analyses or the meteorites from Mars, it had much more iron than our familiar andesites on Earth. There is a great hazard in using such terrestrial rock names on other planets. Andesites are named from the great chain of volcanoes in the Andes. They are major contributors to the growth of our continents. Thus, the presence of 'andesite' on Mars would imply plate tectonics and an earth-like history.

My interpretation is that 'Barnacle Bill' is probably a local product from a basalt volcano, a common enough occurrence on the Earth. So I don't think that there are any lost continents, plate tectonics, or Mt St Helens look-alikes on Mars. The *Pathfinder* Mission landed in a flood channel where rocks from different locations might have been expected. However, the analyses of the soils carried out by *Pathfinder* all look very similar to those from the *Viking* mission at distant locations. This confirms the view that the whole crust seems to be made of basaltic lava, like the ocean floors on the Earth.

To the general astonishment of the scientific community, meteorites from Mars have been found lying on the surface of the Earth. About a dozen are known. Despite early skepticism, we know that they come from Mars. The definitive evidence that they come from the red planet is that they contain some trapped gas that matches that of the thin martian atmosphere that was measured by the *Viking* Landers. The composition of these meteorites is also very similar to that measured on the surface of Mars by the *Viking* Landers.

The random handful of meteorite samples from a geologically complex planet presents us with some difficulties. Suppose we had only a few pieces of the crust of the Earth, and we did not know where they came from. It is also interesting to consider how much we would have learnt about the Moon if all we had were the dozen lunar meteorites that have arrived on Earth. It is easy to argue that one cannot work out the history of a planetary body from a few samples. However, the *Apollo* and *Luna* missions did provide a reasonable sampling of the Moon, since the Moon is well mapped and has had a fairly simple history. Accordingly, it took only a few years of study to establish the history of the Moon. Mars of course is more complex. However, a reasonably consistent picture of the composition and history of Mars is beginning to emerge. Compared with the Earth and Venus, it has accumulated a little more of its share of the volatile elements. Despite this, it has a thin atmosphere. If there ever was an early thicker atmosphere, it has been swept away during great collisions.

A large bulge

Among its other strange features, Mars has a large bulge, which we call Tharsis, on one side of the planet. This bulge is a unique feature among the terrestrial planets. It is about 10 kilometres high at the centre and 8000 kilometres across. It covers about one quarter of the martian surface. The largest volcanoes are sitting on top on it. The great Valles Marineris canyon system, which is really a great rift, starts near the centre of the Tharsis plateau and extends eastwards.

Unlike the Earth, Mars has a rigid crust. There is no sliding around of the crustal plates that we are familiar with on the Earth and that we call plate tectonics. So lavas coming from inside the planet build up great plateaux and volcanoes. These are anchored to one spot. In earlier times, volcanic activity produced basaltic lavas over most of the planet. With time, this volcanic activity has become concentrated in the Tharsis region. For the past two billion years, most lavas on Mars have been erupted in that area, resulting in the great bulge. Why this volcanic activity was so localised must be due to some differences deep inside Mars that we can only guess at.

The great bulge of Tharsis puts the planet off balance. The rotation axis of the planet wobbles about every few million years, tilting between zero and 60 degrees. The consequences for the climate on Mars must be devastating. As the poles move, the polar water ice cap will melt. At present, the tilt is about 25 degrees, but the current similarity to that of the Earth is a coincidence. Life would be more difficult here if we had the large swings in climate that would happen with a greater angle of tilt.

An early wet Mars?

Probably Mars formed an early atmosphere, but being near to Jupiter, it was repeatedly hit by the careering asteroids that the giant was tossing around. Any atmosphere was probably soon knocked off. Being small didn't help either, for tiny Mars had difficulty holding on to any left-over atmosphere.

However, there's more to the story. On the oldest parts of the crust, among the many craters that tell of its ancient age, are some valleys that look somewhat like river valleys on Earth (they are not Lowell's canals) (see Figure 26). So did it rain on Mars in that remote epoch? Because water is needed for life as we know it, the question has excited a lot of interest. Whatever its source, free water cannot exist at present on Mars. Most of the water on Mars is now locked beneath the surface as permafrost. The usual solution is to call on an early 'greenhouse' to provide a thicker warmer atmosphere, from which rain might fall and cut the valleys. Perhaps the temperature could be raised by a large amount of carbon dioxide in the atmosphere. This attractive idea of an early wet and warm Mars, which

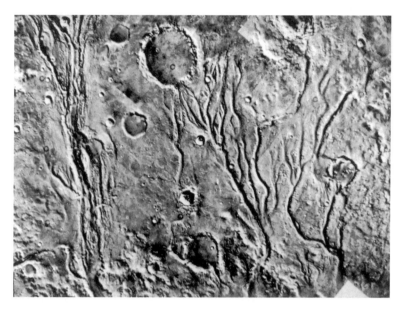

Figure 26. Small river channels on Mars, probably cut by water seeping out from underground rather than by rainfall. These so-called 'valley networks' are in the old heavily cratered terrain in the southern hemisphere of Mars. North is to the lower left and the view is 200 kilometres across (NASA *Viking Orbiter* 63AO9).

would be a bit like the Earth, has run into some fatal objections. The white clouds of carbon dioxide ice will reflect too much of the incoming sunlight to bring the surface temperature above freezing. The problem is even more severe because the Sun was less bright at that distant time. This is the 'faint early Sun' problem that I talk about later.

So what cut the valleys? There is general agreement that they are river valleys, unlike the rilles on the moon that were lava channels. Water could have been trapped in the early crust, like permafrost, where heating by volcanic activity could melt it. As it seeped out, it would cut valleys by headwater sapping, a process familiar on Earth where similar shaped valleys form by water oozing out in springs and undercutting the rocks. The valley heads on Mars are rounded and about a kilometre across. The surfaces between the valleys are smooth and uneroded. Both features argue for an origin by headwater sapping, rather than by erosion by rainfall. The streams can run for a while before freezing even in the thin atmosphere. So probably Mars never had an early Earth-like climate, but was always a cold desert.

Catastrophic floods

However, water has flowed across the surface on a scale that would make the Mississippi look like a trickle. The most dramatic evidence on the Martian surface is the presence of the large channels (see Figure 27). They were detected by spacecraft photos, and have nothing to do with Lowell's canals of last century. These large channels run for hundreds of kilometres. They have streamlined walls. In the midst of the channels there are islands shaped like large teardrops. Such channels can be cut only by massive floods on a scale that Noah might have recognised. The channels appear to start instantly in the

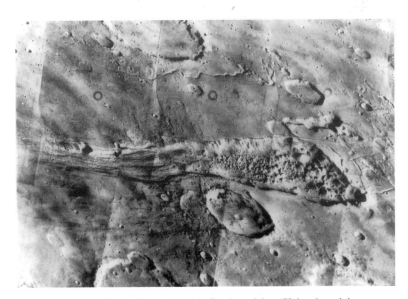

Figure 27. Evidence for catastrophic floods on Mars. Volcanic activity or a meteorite impact has released water trapped beneath the surface as permafrost. The water flooded out leaving the jumbled collapsed terrain that is about 40 kilometres wide and carved out the smooth channels, 20 kilometres wide to the right. The flood had many times the volume of the present Amazon river (NASA S.76–27776).

midst of the volcanic terrains. Probably volcanic heat, or a meteorite impact, suddenly melted a lot of subsurface ice. The water rushed out as a catastrophic flood that reached many times the volume of the Amazon River on Earth.

We have some other evidence that water, trapped as permafrost, is lying beneath the surface. The craters formed by the impact of meteorites are surrounded by sheets of thrown-out debris. In contrast to the boulder-strewn rubble that surround craters on the Moon, those on Mars look slushy, like the splash from a pebble dropped into mud. The heat of the impact presumably melted the subsurface ice and a muddy mix of rock and water was thrown out.

A survivor

Mars is a survivor. While Mars was forming, it was mostly a matter of the poor getting poorer in that neighbourhood. Mars was stunted in its growth, having to make use whatever scraps Jupiter left. The region was in only a little better shape than that of the asteroid belt. That region was left with about one part in a thousand of the material that was originally present in the solar nebula. The rest was tossed away by the neighbouring giant. Mars was slightly better off, finishing up only a bit over a tenth of the mass of the Earth.

Other Mars-sized bodies formed within the inner solar system, but did not survive. They were swept up by Venus or by the Earth. The impact of one such body provided us with the Moon, of which more later.

4

The twins

Venus and the Earth are close in size and composition. These planets should be look-alikes. Here I investigate the differences that resulted when nature tried to make twin planets in the solar system.

Venus

The evening star

Venus, rising in the morning or setting in the evening skies, is the most brilliant object in the sky, after the Sun and the Moon. This gleaming jewel has been admired since antiquity. Because it is an apparent twin of the Earth, it has always been of intense interest as the only similar planet in the solar system. When Venus was found to have an atmosphere, it did not take much imagination to make it a hotter version of the Earth. It was soon clothed with thick tropical forests and swamps that were populated with various monsters. Dinosaur-like creatures were favourites of science fiction writers. The coal-making swamps of the Carboniferous Period on the Earth, complete with giant dragonflies and exotic trees, provided other likely models for fevered minds. This idea of a tropical version of the Earth made sense as Venus was closer to the Sun.

Dr Jekyll and Mr Hyde

Venus is a little smaller in radius than the Earth, but has the same density, when allowance is made for the small size difference. So there is probably not any real difference in bulk composition. The apparently trivial difference in size turns out to be one of the crucial factors in making Venus different to the Earth. In contrast to the hopes of science fiction writers, Venus only looks like the Earth in the same sense that the evil Mr Hyde resembled the good Dr Jekyll in the novel by Robert Louis Stevenson (1850–1894). Early observers thought that the planet was either spinning rapidly, similar to the 24 hour period of the Earth, or perhaps on a monthly period. However, a surprise was in store. When radar in the 1960s penetrated the massive cloud cover, Venus was discovered to rotate very slowly backwards. Unique among the planets, it takes 243 days to make one rotation, although the atmosphere of Venus rotates in about four days. Venus orbits around the Sun in only 225 days. Thus, the day on Venus is longer than the year. Unlike the Earth, Venus has no moon. The planet has an atmosphere, mostly of carbon dioxide, that is about 50 times denser than our own. The atmospheric pressure is 90 times that of the Earth. There is only a trace of water vapour in the atmosphere, and none, of course, on the surface, where the temperature is 477 degrees centigrade (750 K), twice as hot as a kitchen oven, and hot enough to melt lead. Whether there is water trapped deep in the interior is an open question. It is possible that the planet is almost totally dry. To complete this litany of differences, Venus has no detectable magnetic field. A navigator would be hard-pressed to find his way, as the stars would not be visible through the thick atmosphere either.

There is thus a certain irony that the surface of Venus was revealed in stunning clarity by the radar on the *Magellan* spacecraft, named for the celebrated explorer. The landscape that it has uncovered must inspire a sense of humility, since it is so different from that of the Earth. No hard-won models of geological behaviour, worked out from long and diligent study of the surface of the Earth, are of any use on Venus. We have to start anew. This is a common tale throughout the solar system. Each newly explored planet and satellite has some variation from our knowledge gained by studying our own planet.

The Earth's crust is divided into a thick low density continental crust and a thin dense crust of basalt that underlies the oceans. Like a baker who can only make one loaf, Venus seems to have produced only a monotonous crust of basalt. Thus, there are remarkable secondary differences between the two planets, when one considers that they are so similar in density and size.

What are the causes of the differences? Probably many are accidental. Perhaps no big body hit Venus, so that it spins slowly. It may
have no moon for the same reason. Nothing struck the planet at the
right angle to splash one off. The thick atmosphere remains because
nothing big enough to remove it hit the planet. The high surface temperature follows from this. The similarity in density to the Earth and
the obvious presence of basaltic lavas imply a similar internal structure of a metal core and a rocky mantle. The absence of a magnetic
field seems to be due to the slightly smaller size of Venus. It is the
freezing of the Earth's inner core which is thought to drive the dynamo
and so produce our magnetic field. The pressure deep inside Venus
is just a little too low to make a solid inner core. Thus, Venus and the
Earth, although so close in size and uncompressed density, have
evolved very differently. They are not 'twin' planets. So significant
differences between planets can arise from small changes in size.

A one-plate planet

The surface of the Earth is covered by over a dozen large plates that
jostle one another. Venus, in contrast, is a 'one-plate' planet. If the
water and the thin veneer of mud were removed, the basalt-covered
floors of our oceans would look somewhat like the surface of Venus.
However, it is clear that there are several major differences between
the surface of Venus and the lava-covered floors of the oceans. There
is no equivalent on Venus to the great mid-ocean ridges on Earth,
where fresh lava comes to the surface, and from which the sea floor
spreads away. Although there is a little local lateral movement on
Venus, the extensive spreading that is so characteristic of our oceanic
crust does not occur. Nor is there any sign on the surface of Venus of
the great trenches in our oceanic crust. This is the location where the
oceanic crust gets pulled back down and recycled into the mantle. On
Venus, there is nowhere for the lavas to go. Venus has choked itself
with a thick crust of lava that it poured out over the surface.

   Unlike the Earth, the bone-dry rock at the surface of Venus is
very strong. It can hold up steep slopes for millions of years. On the
Earth, the mountains float like icebergs, supported by deep roots. On
Venus, they just sit on the surface. This is strong enough to hold them
up, like the mythological Atlas, who carried the whole Earth on his

shoulders. Some of the mountains on Venus are as high as those on Earth. The volcanic massif of Beta Regio is nearly ten kilometres high. The major mountain range, the Maxwell Montes, rises eleven kilometres above the low rolling plains that cover most of Venus. These high standing areas that might be mistaken for continents are crumpled up basaltic volcanic crust. The strength of the crust on Venus seems to be due to the absence of water. Like an armadillo, Venus has encased itself in a strong dry rigid shell of basalt.

Although Venus must have had a similar store of heat to the Earth, it seems to have splurged it in a wasteful manner. In great contrast, the Earth has rather carefully conserved its energy by recycling the crust under the oceans back down into the mantle, using this process along the way to make our useful continental crust.

The crust of Venus

Nevertheless, some history can be seen on the crust of Venus. When faced with something new, we invent a new jargon as a defence. Giving something a name is often used as a substitute for an explanation. I apologise for introducing some of these exotic terms, but they might help anyone who wants to delve further into the history now revealed of our twin. There are three main regions. The oldest parts are crumpled-up crust called tesserae. Most of the surface is covered with featureless rolling plains of basaltic lavas, typically about 300–500 million years old. These are only a little younger than the tesserae. The youngest features on the surface are volcanoes and some circular structures a few hundred kilometres in diameter called coronae. This is the end of the jargon for the meantime.

There are many craters caused by the impact of comets and asteroids. More on these later. On the volcanic plains, there are thousands of small shield volcanoes, a few kilometres in diameter. Over 50 000 have been identified. They are just sitting on the surface, supported by a thick crust. A few small 'pancake' features, about 20 kilometres across, appear to be composed of more viscous material, which has spread out like treacle. They may be like the domes of viscous lavas on the Earth. These are usually of granitic composition. The 'pancakes' on Venus are isolated and not to be confused with the broad expanses of granite on our familiar continents. Thus, the production of great masses of granite, as we see on our continental crust, does not occur on Venus. This rock, so familiar to us from city buildings and pavements, may be uncommon in the rest of the solar system.

Fresh craters on Venus

The craters formed by meteorite impact on the surface of Venus are surprisingly fresh (see Figure 28). Only rarely are they covered or entered by lava flows. No craters less than five kilometres in diameter have been detected on Venus and there are hardly any craters with diameters less than 30 kilometres. This is a tribute to the thick

Figure 28. A fresh impact crater on Venus, revealed by the *Magellan* radar looking through the thick atmosphere. This is the crater Aurelia, 32 kilometres in diameter, excavated when an asteroid a few kilometres in diameter hit the barren basalt plain, here shown as the dark smooth area surrounding the crater (NASA *Magellan* P-37128).

atmosphere of Venus. The smaller meteorites explode or are burnt up due to friction as the incoming body ploughs through the dense clouds. Our own atmosphere, which is 50 times less dense, is a much weaker shield. However, it protects us from the smaller bits and pieces. We see these 'shooting stars' on clear nights in the country.

When a meteorite or comet hits the surface, debris is thrown out by the explosion and surrounds the crater like a blanket. Unlike craters on other planets, those on Venus often have a missing sector in this blanket. This gap is caused when the rubble thown-out by the explosion runs into the turbulence in the atmosphere caused by the incoming meteorite or asteroid. The flung-out rock just gets tossed aside. There are also numerous dark smooth patches or 'splotches' extending over many kilometres. These are apparently caused by shock waves which blast and scour the surface. These result from meteorites that are too small to penetrate through to the surface and which explode high in the atmosphere. The Earth had a similar experience in 1908, when a meteorite about 60 metres in size exploded five kilometres above the Tunguska River in Siberia. The blast blew down the forest for up to 30 kilometres distant.

Multiple craters are common on Venus. These are caused when the incoming projectiles break into several pieces in the atmosphere. In striking contrast to the lunar craters, the beautiful far-flung ray systems are missing. This is again a consequence of the thick atmosphere, which stops the spray of fine particles from the impact.

The youthful face of the goddess

It seems appropriate that the face of the planet named for the goddess of love should have a young face. The age of the surface of Venus can be worked out from the number of craters that have formed on it over time. We know the rate of impacts from our study of the craters on the dated surfaces on the Moon. Based on this, the surface of Venus appears to be relatively young. This is a very strange finding. On Earth we are missing only the first ten per cent of our history in the rock record, but on Venus, there is no record of what happened in the first 85 per cent of its history. There are no old surfaces covered with craters, such as we see on Mars, Mercury or the Moon. The present surface of Venus is somewhere between three and five hundred million years old. What is curious is that not much seems to have happened there since. The 950 impact craters that were produced on Venus over that period of time are mostly uneroded. Five hundred million years ago, the Cambrian Period on the Earth was marked by the appearance of the first hard-shelled animals. Quite a lot has happened here since then. Venus has had a very different history. A few hundred million years ago, it covered its surface with a great splurge of lava. Apparently exhausted by this great outpouring, geological activity on Venus hasn't managed to produce more than a trickle of lava since then.

The most sobering aspect of the craters on Venus is that there are 950 of them on a surface that is only a few hundred million years old. This means that an impact forms a large crater every half million years. Venus is close to the Earth in size and location. We present a similar target to this bombardment. This provides us with the chilling information that a large impact crater must also have formed on the Earth about every half million years. Most of this record has been erased by erosion on our planet.

Four of the craters we see on Venus are larger than 130 kilometres in diameter. The largest is Mead, named after Margaret, the anthropologist (1901–1978). It is 269 kilometres in diameter. These craters are ringed basins of the sort that has been found buried under sediments at Chicxulub, on the Yucatan Peninsula in Mexico. That one was responsible for the extinction of the dinosaurs and much else.

Water on Venus?

The surface temperature of the planet is far above the boiling point of water, but the atmosphere contains only a trace of water vapour. Did Venus ever possess more water? Were there early oceans? Because the Earth and Venus are commonly thought of as twin planets, it is natural to think of an early Venus covered with an ocean that vanished due to overheating by a 'greenhouse' effect. This is a current cause for alarm that something similar will happen to the Earth if we put too much carbon dioxide into the atmosphere by burning up fossil fuels. Back in the Cretaceous Period (The Age of Chalk) when the dinosaurs roamed, the Earth had about three times as much carbon dioxide in the atmosphere. The climate was much more pleasant than what we experience in the unstable interglacial period in which we live and which we regard as the norm.

My own perception is that Venus never collected much water to begin with. The loss of surface water on Venus by the greenhouse effect does not explain why the planet is so thoroughly dry. It's very difficult to remove the last few per cent of water, even from a surface as hot as that of Venus, mainly because of the presence of a cold trap in the high atmosphere, which would condense the water as the temperature falls. It has been suggested that some of the water might be trapped in the mantle if some kind of plate tectonic process was operating. This seems unlikely on Venus. The problems of how much water, if any, that Venus had originally, and what happened to it, remain as current problems.

A close relative of the Earth?

It is sobering to contemplate the differences between those apparent twins, Earth and Venus. The differences seen in the geology of the surface, the absence of plate tectonics on Venus and the different rates of volcanic activity are probably ultimately due to the differences in the amount of water in the planets. All these features are the result of chance events during the formation and evolution of the planet. They do not encourage the speculation that earth-like planets are common. None of the planets in the solar system nor the 60 or so satellites resemble one another; all are different and could be members of another planetary system without exciting comment. The message is that chance events have played a crucial role in the origin and evolution of the solar system and that planets similar to the Earth are unlikely in other planetary systems. It is the Earth that turns out to be the exceptional planet, not Venus.

The Earth

An island entire of itself

In many ways the problem of accounting for the Earth resembles that of the origin and evolution of the solar system; only one example of each exists. How can one deduce general principles about the solar system from a unique planet accompanied by a unique satellite? A good example has been the difficulty in recognising on the Earth that the formation of craters by the impact of asteroids, comets and meteorites is an important planetary process. Erosion has removed most of the evidence. It is only recently that professional geologists have come to accept the possibility of such a continuing bombardment. Our experience with lunar geology and geochemistry, with its subtle but crucial distinctions from our hard-won experience on the Earth, should also remind us of the hazards of trying to extrapolate from unique terrestrial conditions. Possibly it is better to remind ourselves about some regions of the interior of the Earth of which we have little knowledge, for it is ironic that we understand the composition and evolution of the Moon better than that of the Earth.

The great philosophical contribution made by the study of the Earth was to establish the immensity of time, which is illustrated by almost any geological feature (see Figure 29). The 6000 years that Archbishop Ussher calculated have been replaced with an age for the Earth of four and a half billion years, itself only a fraction of the age of the rest of the universe.

So what does one say about this planet, which is unique even by the standards of the solar system? We know so much detail about the Earth that any attempt to summarise our knowledge risks reducing the discussion to a trivial level or to a recitation of truisms that everyone knows. It is also useful to remind ourselves that one of the great obstacles to progress is 'not ignorance, but the illusion of knowledge'.[2] Here I concentrate on the early stages. Once I have described how the Earth was formed, I reluctantly have to leave most of rest of the wonderful saga of the geological history of the Earth to others[3] and press on with the main thrust of this book.

The composition of the Earth

However, a few areas remain that are profitable to discuss in the present context. The first is the bulk composition of the Earth. How does it compare with that of the other planets? One of my colleagues remarked that

> estimating the compositions of planets has proved to be one
> of the most difficult tasks in cosmochemistry. Even determining
> the composition of the Earth is a challenge, because planetary

one metre

Figure 29. The immensity of geological time is well shown in this picture
of the famous unconformity at Siccar Point, near St Abb's Head on the
east coast of Scotland, north of Berwick on Tweed. The flat-lying rocks on
top are the Old Red Sandstone, which has been uplifted and tilted about
15 degrees. The vertical strata underneath them had long before been
sands and muds in horizontal beds, before being raised, tilted, and planed
off by erosion. Then they formed the base on which the sands of the Old
Red Sandstone were deposited in the Devonian Period (in popular language, the Age of the armoured fishes). The underlying vertical strata are
Silurian in age, so that this particular example records the passage of about
50 million years. Although this is inconceivably long on human time scales,
it constitutes only about one per cent of geological time. Such examples
abound in the geological record. John Playfair (1748–1819) wrote on visiting this outcrop about 1790, 'We felt ourselves . . . carried back to the
time when . . . the sandstone before us was only beginning to be deposited,
in the shape of sand and mud, from the waters of an . . . ocean . . . An epoch
still more remote presented itself, when even the most ancient of these
rocks, instead of standing upright in vertical beds, lay in horizontal planes
at the bottom of the sea . . . revolutions still more remote appeared in the
distance of this extraordinary perspective. The mind seemed to grow
giddy by looking so far into the abyss of time'.[1]

differentiation has ensured that there is no place on or within it where we can collect a sample that has the bulk composition of the whole planet.[4]

The abundances of the chemical elements in the primitive meteorites give us the best estimate of the composition of the rocky component of the solar nebula. One might have thought that the composition of the rocky planets like the Earth would match this. Many theories of planetary formation have indeed supposed so. However, the terrestrial planets have lost not only the gas and ice but are also depleted in elements such as lead, sodium and potassium that are volatile at temperatures less than 1000 K. This was because the early Sun swept these elements away, along with water, the other ices and gas during its violent youth.

A major problem is that there are two substantial regions in the Earth about which we have little definite information. These are the lower half of the continental crust and the lower parts of the rocky mantle. Both are important, but represent unknowns in attempting to establish the overall composition of the Earth. That of the upper crust is quite well known. After all we live on it and geologists have been hammering, probing and drilling it for a couple of hundred years.

The crust under the oceans, which covers about three quarters of the surface of the Earth, is mainly basalt, about five kilometres thick, which has erupted as lava from the mid-ocean ridges, and which is covered with a thin veneer of mud that comes from the erosion of the continents. The composition of the rocky upper mantle is also relatively well understood down to about 200 kilometres, as we have samples from it brought up by volcanoes. Deeper down, our knowledge is less certain, although we know from studying the passage of earthquake waves that the mantle is still rocky. The core, which begins at 2900 kilometres below the surface, is mostly metallic iron, alloyed with about ten per cent of nickel, with sulphur and maybe some other elements.

There are continuing controversies among experts in many of these areas. Although the meteorites provide us with important information, what is clear is that the Earth has not been put together from a mixture of the meteorites that we can study. This is a common fallacy. The meteorites come from much further out in the solar system and have significant differences in composition from the Earth such as the proportions of noble gases, such as neon and xenon, as well as the proportions of the more common elements.

On reflection, the asteroid belt is not a very good quarry from which to build the Earth. The early formation of Jupiter managed to get rid of nearly all the material there and left a hole in the nebula with less than one part in a thousand of the mass of the Earth. Of course one also has to find material from which to build Mercury, Mars and Venus. Clearly, one has to look sunwards from the asteroid belt for the material that built up these planets. The meteorites remain a valuable, if depleted, source of information, just as refugees can tell stories of their homeland.

The upper rocky mantle of the Earth has too much magnesium, among other elements, to come from any meteorite that we know about. This observation has been the focus of a continuing argument. The unknown composition of the deeper mantle is a decisive factor. If it has the same composition as the upper mantle, which I think likely, then the Earth has a composition different to the primitive abundances of the chemical elements. So we are still having problems with the home planet. It seems clear that that the Earth must have accreted from a local region and that there was not much mixing from as far away as the asteroid belt during the formation of the Earth.

The accumulation of the Earth

What were the sizes of the bodies that formed the Earth. We can only study this problem by computer models, as the originals have vanished, swept up into the planets. These models inform us that between half and three quarters of the Earth was put together from bodies the size of the Moon or larger. One was perhaps a quarter of the mass of the Earth. These lost bodies would have been respectable planets, bigger than Mars or Mercury, if they had survived.

The largest of them most probably arrived during the closing stages of the accumulation of the planet about 4500 million years ago. Some left-over objects were swept up in the next 500 million years. The clean-up of these bodies was mostly completed by about 3850 million years ago. We know this time rather precisely from our study of the samples from the Moon. Besides many smaller craters, over 200 great basins, bigger than France, formed on the Earth during this period. This bombardment would have destroyed any early crust. The smashed up rock would have been easily removed by erosion, unlike on the Moon, where the cratered highland crust is still preserved. This continuing bombardment explains why there are no rocks on this planet that date back to the beginning of the solar system. This is not a good planet on which to try to understand what happened in the earliest times.

High temperatures are a consequence of this method of forming the Earth. Total melting of the Earth seems unavoidable. The metal core of the Earth formed quickly. Being heavy, the molten metal just fell in to the centre, like iron in a blast furnace. This contrasts with the great gas giant Jupiter, over 300 times more massive. In that planet, as we have seen, its core had to form first. The inner part of our core has now cooled and is solid, but the outer part of the iron core is still molten. This knowledge about the deep interior of the Earth is well established from studies of the transmission of earthquake waves. The rocky mantle cooled and became solid very quickly, but just how this happened is a bit of a mystery. We understand very well how this occurred on the Moon, but the sheer size of the Earth has defeated the attempts of modellers to understand what happened here and there are few clues yet. Probably it cooled and quenched very quickly, perhaps within 1000 years, before the crystals had a chance to separate. Certainly there is no sign of the sort of crystal sorting that occurs in small pools of melted rock on the Earth, and that is well understood by geologists. A similar process that formed different zones of minerals also occurred on a larger scale during the cooling and crystallisation of the Moon from its initially molten state.

Adding icing to the cake

The rocky mantle of the Earth has more nickel, platinum, iridium and similar precious elements than expected on simple models. By rights these elements should be all in the core along with the iron, and we would have no platinum wedding rings. So the idea is that a veneer of meteorites was sprinkled in late in the formation of the Earth. Adding this veneer to planets that are essentially complete is like adding icing to a cake. The decoration on top may give little information about the composition of the interior.

However, the Moon shows no evidence of such a late veneer, although it should be obvious if it were there. It is also worth making the point that if there were late veneers they must have been thoroughly mixed into the rocky mantle, which was probably molten at the time. But perhaps we just don't understand enough about the behaviour of the chemical elements at the high temperatures and pressures deep inside the Earth. Maybe the problem of the presence of the precious elements is our limited understanding. Perhaps the giant Moon-forming impact event that I discuss later was the culprit that added these elements. In that model, the metallic core of the impactor, after being in orbit for a few hours, fell into the Earth. As it dumped its mass of metal into the Earth, some of it may have been trapped in the rocky mantle.

Crusts

Planets, like bakers, seem unable to resist making crusts. In both cases heat is involved. Easily melted bits rise from the interior and coat the surface. However, the solid crusts of the planets and satellites all differ from one another. Accordingly, there are difficulties in trying to discover some general patterns for the formation of crusts in a planetary system in which random events are common. There is clear evidence that the Moon was melted and formed its white highland crust as a result. However, this does not necessarily provide us with a model for the development of crusts on the early Earth, Venus or Mars.

The familiar continental crust of the Earth that most of us live on is of unique importance because it formed the platform above sea level on which the later stages of evolution occurred. These led to the appearance of *Homo sapiens* and to the writing of this account. It is thus of interest to enquire how the continental crust, which differs so greatly in composition from the oceanic crust and the mantle, came into existence, and whether similar crusts exist on the other terrestrial planets.

Familiarity with our own crust perhaps has obscured how remarkable crusts are. Although the continental crust, about forty kilometres thick, is less than half of one per cent of the mass of the

Earth, it contains a surprising amount, about one third of the total Earth budget, of many elements present in trace quantities such as uranium.

Occasionally, geological processes conspire to concentrate these elements, normally present at a few parts per million, or billion, into ore deposits. So the crust is rich in mineral deposits that contain these scarce elements that we find so useful in building a technical civilisation. Every colour television screen contains the rare earth element, europium, which fluoresces to provide the red part of the image. Europium is present in the Earth at a level of only one tenth of a part per million, and rarely (hence the term 'rare earth') occurs in minable amounts. How many other planets might reproduce the endless geological cycles that have concentrated this element. Here we can extract enough of it from exotic mineral deposits (costing even so about $8000 per kilogram) to spread it around the world in television sets. So for a technical civilization to develop, one not only needs a favourable temperature, water and an atmosphere of oxygen. One also needs deposits of copper, rare earths and lots of other things that we take for granted, but come to us courtesy of plate tectonics. In contrast, the crust of Venus looks like a prospector's nightmare.

A basic question on Earth is what happened in the gap of about 500 million years before the formation of the Earth and that of the oldest rocks that we recognise. We can place a few limits. There is no evidence on the Earth for that enduring myth of geology, a primitive world-encircling crust of granite. The lack of evidence for such a crust is overwhelming. Our common granite, which decorates so many city buildings, turns out to be difficult for planets to make, so that the production of a granitic crust is probably unique to the Earth. It is the end product of three or more stages of distillation from the primitive rocky mantle. The continents have grown slowly and in an episodic manner throughout geological time and are now at their greatest extent. This process of forming the terrestrial continents is clearly inefficient. The Earth has transformed less than half a per cent of its volume to continental crust of intermediate composition and less than one fifth of one per cent of its volume into the upper granitic continental crust in over 4000 million years. No company in the business of manufacturing continents would stay in business with that sort of record.

The turbulent history of the atmosphere

Readers should be aware that more problems than answers are to be found. Certainly, the atmosphere of the Earth has been through a series of events of staggering complexity. There appears to be little trace of any primitive gases. If the Earth formed within the gas-rich disk, it would have captured a thick primitive atmosphere. The 'rare' gases, like neon, would be a hundred times greater than is their present atmospheric content. The low abundance of the rare gases in our present atmosphere does not fit this model by very wide margins. Accordingly, it appears that the gases had gone by the time the Earth got around to forming. As a final insult, large collisions while the Earth was being put together would have stripped away any primitive atmosphere. Thus, the present atmosphere and oceans of the Earth appear to be entirely secondary in origin, and so provide little information relevant to this inquiry.

However, the nature of the early atmosphere is of much interest with respect to the origin of life. It is not known whether the atmosphere was reducing or non-reducing, but it does not appear to have been strongly reducing. Free oxygen was absent and did not become available until a little before 2000 million years ago, when photosynthetic bacteria capable of producing oxygen became abundant.

The scarcity of water

In contrast to the other inner planets, the significant feature about the Earth is the presence of liquid water at the surface. This is a critical factor both in facilitating the process of plate tectonics and in allowing the oceanic crust to be recycled through the mantle. The water trapped in the crust beneath the ocean eventually is taken down deep into the mantle. There it plays a critical role. As the crust sinks down into hotter regions, the water is driven off, taking with it many of the more volatile elements. As this fluid rises, it triggers melting within the mantle. This lava, now full of volatile elements, reaches the surface in spectacular fashion. The eruption of Mount St Helens was a typical example. These events are seen most dramatically in the great chains of explosive volcanoes that occur around the rim of the Pacific Ocean, at the locations where the oceanic crust slides back down into the mantle. Such processes supply the material that go to make up the continents, along with our useful ore deposits. In other planets the basaltic lavas stay on the surface. This leads to the persistence of barren basaltic plains such as those that we observe on Venus, Mars and the Moon.

Everyone is impressed by the great abundance of water on the surface of the Earth. However, it is worth recalling that the amount of water on the Earth is very small in cosmic terms. The solar nebula in the neighbourhood of the Earth became as dry as the Sahara as water and other volatiles were swept out by the early active Sun to where Jupiter was forming. If the Earth had its proper share of what was present in the original disk of dust and gas, it would have nearly a 1000 times more water. This amount would treble the volume of the Earth, and drown us in in a deluge that would have astounded Noah.

So if water is so scarce in the inner solar system, where did the streams, lakes, rivers and oceans that we admire come from? Their source is an interesting question, not yet fully understood. Surprisingly, this could be treated as a trivial problem, since the Earth currently contains only about 500 ppm water. This is such a small amount that the source of the water on the Earth could be ignored, except that we are here as a consequence on this 'bank and shoal of time'.[5]

Comets are a possible source. However, one major difficulty is that the deuterium/hydrogen ratio in cometary water is too high for them to be major source of water for our oceans. Comets, coming from far away, mostly arrive at higher velocities than meteorites. Thus they may take away more than they bring and like an unreliable delivery man, they may remove atmospheres as readily as they deliver them. However, they are probably the ultimate, if fickle, source for rare gases, organic compounds and atmospheres.

Is the Earth alive?

I cannot conclude a discussion of the Earth without some reference to the entertaining hypothesis of James Lovelock (b. 1919) that the Earth is alive. This is discussed by him in several books dealing with the so-called Gaia hypothesis. Gaia was the Earth Goddess in Greek mythology. There are 'strong' and 'weak' versions of this notion, somewhat like the versions of the anthropic principle that I discuss later. In the 'strong' version the Earth is a sort of superorganism. In the 'weak' version, life influences the environment by various feedback processes, and strives to maintain a suitable environment.

According to Lovelock,[6] the Gaia hypothesis states that

> the atmosphere, the oceans, the climate, and the crust of the Earth are regulated at a state comfortable for life because of the behaviour of living organisms. Specifically the temperature, oxidation state, acidity and certain aspects of the rocks and waters are at any time kept constant, and this is maintained by active feedback processes operated automatically and unconsciously by the biota.

Two examples of this line of reasoning must suffice. Firstly, changes in solar radiation would be consciously accommodated by a greater or lesser biological production of greenhouse gases to maintain an even surface temperature.

A second example comes from the oceans. In order to keep the oceans from becoming too salty, and therefore a problem for life, the proponents of Gaia suppose that reef-building organisms (such as corals and stromatolites) construct off-shore reefs. The lagoons shoreward of the reefs provide a location where sea water can evaporate, and salt deposits form, thus removing salt from the oceans by trapping it in geological strata. According to Lovelock, 'the balance of erosion and formation always seems to have kept enough salt sequestered in evaporite beds to keep the oceans fresh and fit for life'.[7]

The basic question is whether life is controlling the environment, or whether life is struggling to survive in the face of changes imposed by geology. Life actually seems very well adapted to such a purpose, as shown by the rapid reintroduction of life into regions devastated by a catastrophe. The rapid greening of Mount St Helens is an example. After the devastation caused by the eruption and massive sideways collapse of the volcano, only a few years were needed to restore life into many niches. Life acts on very short time scales in comparison with the enormous periods required for major geological evolution. Short-term local catastrophes such as hurricanes, earthquakes, tidal waves (tsunamis) and volcanic eruptions, all of which are traumatic for those involved, have little long-term effect.

The Gaia proposal is difficult to distinguish from the general perception that life is exceedingly well adapted to the environment, with every niche filled. In this view, life would presumably have adapted to a saltier or fresher ocean, or to a different temperature. When one considers the range of temperatures that life tolerates, from boiling hot springs to snowy Arctic wastes, one is struck by the adaptability of life to external circumstance.

How the Gaia hypothesis can accommodate the sort of chance events that dominate this book is not clear. Catastrophic meteorite or cometary impacts, and even ice ages and major volcanic eruptions all seem to pose insurmountable obstacles by having the potential to overwhelm life, perhaps on a global scale. It becomes a matter of bad luck. No matter how carefully the coral builds its reef, it has cannot anticipate the arrival of the asteroid on its accidental orbit. Thus, many coral species were lost when life on Earth came close to extinction 250 million years ago, in the great, but little understood, catastrophe at the Permian–Triassic boundary, an event that extinguished 95 per cent of all life on the planet.

It seems very difficult to propose tests of the Gaia hypothesis. One suggested test is that if life had developed early on another planet and subsequently became extinct, this might disprove the hypothesis. Perhaps we will have an answer to this question, since it is possible that primitive bacteria arose on Mars and died out. But the extinction of life, as that of a species, might just as easily turn on the bad luck of a meteorite strike. Since neither version of the Gaia hypothesis seems to make predictions that appear to be easily testable, the notion remains at present as an intellectual curiosity, in much the same situation as the anthropic principle that I talk about shortly. Neither hypothesis is compatible with the view expressed in this book that chance events have dominated both the development of the planets and the evolution of life.

5

Two special cases

The Moon and Mercury are special cases even by the rather loose standards of the solar system. Mercury is unique on account of its high density that tells us that it has a high content of metallic iron relative to rock. In contrast, the Moon has the reverse, a low content of metal relative to rock. Explanations for the peculiar nature of both bodies have a long history and much effort has been expended in attempts to fit one or both into overall schemes of planetary formation, but without conspicuous success until recently.

The Moon

An eccentric individual

The Moon is the most obvious object in the sky, except for the Sun. The dark areas that make the features of 'The Man in the Moon' are obvious to everyone, while a low power telescope or even binoculars reveals the craters that so impressed Galileo four centuries ago. He saw that the Moon was mountainous, thus fulfilling the prediction of Anaxagoras (*c.* 500–428 BC) that the Moon was stony. History tells us that truth is often dangerous and the philosopher was banished from Athens for this heresy. Galileo had similar problems.

Laplace commented ironically that it was an ancient view that 'the Moon was given to the Earth to afford it light in the absence of the Sun',[1] although due to some oversight the Moon shines for only half the time that it is needed. The beauty of moonlit nights has often been celebrated. Thus, it is not surprising that the Moon has had a powerful, as well as a romantic, effect on human development. The other planets and stars are so distant as to be points of light in all but the most powerful telescopes. It thus takes an extraordinary feat of imagination to understand their true nature, but the surface of the Moon tells us of the existence of other worlds like the Earth. Without the Moon, our intellectual development would have taken a different, probably more inward-looking course. The waxing and waning of the Moon on a monthly cycle provided primitive calendars. Curiously, the lunar month is the same length as the human female menstrual cycle. Human female pregnancy lasts for nine lunar months. No one understands the significance, if any, of these coincidences. Perhaps they account for our romantic attachment to the Moon, which has long been applauded in song and poetry. The connection of the Moon with tides was realised at a very early stage, and eventually led to the idea of gravity, developed mainly by Descartes and Newton.

The Greek cosmologists had proposed that the heavenly bodies, unlike the Earth, were perfect and made of shining crystal, the fifth element or quintessence. However, Galileo's observations showed clearly that the Moon bore some resemblance to the Earth. He was bold enough to distinguish 'land' (the white lunar crust) and 'seas' (the dark lava flows). This marks the beginning of the comparative study of the planets.

The Earth–Moon system is unmatched among the inner or terrestrial planets. Neither Venus, close in size to the Earth, nor stunted Mercury have moons. Phobos and Deimos, the two moons of Mars, are probably tiny captured asteroids. Relative to the Earth, the Moon has the largest mass of any satellite in a satellite–planet system (1/81). The satellites of the outer planets are mostly ice–rock mixtures, whereas the Moon is made of rock. The Earth–Moon system is spinning much more rapidly compared with other planets. Something kicked the Earth. Curiously, the lunar orbit is not in the plane of the Earth's equator, nor in the Earth–Sun plane, but is inclined at five degrees to the latter.

Rosetta Stones

The Moon has played a central role in the development of theories of the origin and evolution of the solar system. This is not without irony, since it has proven one of the most difficult objects to explain. It is in plain sight, accessible even to naked eye observation, as Harold Urey (1893-1981), who persuaded NASA to go to the Moon, was accustomed to remind us. It is an obvious first object to fit into theories for the origin of the universe. The Moon was often thought to be a kind of Rosetta Stone,[2] so that the general belief in pre-*Apollo* days was that we could discover much about the origin of the solar system by going to the Moon. This was a major scientific justification for the manned lunar missions, although political considerations were the real driver.

The Moon eventually provided us with a kind of Rosetta Stone for understanding the past history of the solar system, but not in the manner imagined by Earth-bound thinkers before the *Apollo* missions. One of the principal conclusions of lunar studies was to demonstrate the importance of large impacts of asteroids, meteorites and comets. The evidence from the wide range of impact crater sizes led to the notion that a variety of objects of differing sizes existed and that the planets formed from these rather than from dust. An understanding of the importance of massive impacts early in solar system history enables us to comprehend the origins of both bodies, as well as many other features of the solar system such as the tilts of the planets.

Richard Dawkins (b. 1941), the biologist and modern defender of Darwin's 'dangerous idea', has pronounced that 'the Moon is simple'.[3] Indeed, it might appear so to a biologist attempting to account for the development of the eye, but the nature, composition, evolution and origin of the Moon baffled scientists until very recently. Before the lunar missions, the heroic efforts of cosmologists and a wide variety of experimental scientists all failed to provide an adequate explanation for either the composition, the history or the existence of the Moon. It is sobering to ponder the attempts to explain the origin of the Moon and planets before 1969, which involved so many false trails.

With the benefit of hindsight, it is clear, however, that several key facts, among them the low density of the Moon, the strange orbit, and the rapid spin of the Earth–Moon system, were available already, awaiting integration into a coherent theory

The space missions provided crucial additional information on ages, chemistry, and the significance of cratering, in particular the importance of large basin-forming impacts. The largest structures formed by such processes are basins surrounded by circular rings of mountains. The type example is Mare Orientale, which is the size of France. It formed in a few minutes when a body of the order of 50 kilometres in diameter travelling at many kilometres per second, crashed into the Moon (see Figure 11).

Many features of the solar system, such as the various tilts of the planets, could be explained by such random processes, so that a new understanding of the origin of planets and satellites emerged. Once we properly understood the origin of the Moon with all its implications and connotations, then a clearer view of the rest of the solar system began to emerge, and the somewhat untidy nature of the solar system received a rational explanation.

A thick crust

The surface of the Moon is covered with a blanket a few metres thick of rubble and dust from the impacts of meteorites, so that it presents a rounded rolling surface to the astronaut. The absence of familiar landmarks makes it extraordinarily difficult to judge distances. There is a surprising amount of relief, over 16 kilometres between the highest and lowest point. This is only a little less than the extremes of 20 kilometres on the Earth between the top of Mt Everest in the Himalayas and the Challenger Deep in the Marianas Trench of the western Pacific Ocean. On the Moon, the rugged terrain is the result of giant impacts, gouging great basins, rather than the forces of plate tectonics that continue to shape the surface of the Earth.

The Moon has a thick crust, about 12 per cent of planetary volume, which formed quickly following the formation of the Moon at about 4.45 billion years ago. The crust varies between 60 and 100 kilometres thick, on a body whose radius is only 1738 kilometres. The continental crust of the Earth, in contrast, has grown slowly in fits and starts throughout geological time. Our crust is relatively much smaller, comprising less than half a per cent of the volume of the Earth.

The lunar highland crust is different in composition from the interior and contains a large proportion of feldspar, which is responsible for the white colour of the lunar highlands. It is complex in detail, mainly because the rocks were smashed up by the meteoritic bombardment. However, the chemical composition survived to tell its tale. The generally accepted model is that the crust formed as crystals of feldspar floated like ice-floes on the surface of a molten Moon.

The structure of the lunar crust was dominated by the early large basin-forming impacts that produced circular rings of mountains. The fact that the mountain ranges on the Moon lay along arcs of circles was a great puzzle to early investigators. Now we understand that they form like giant ripples of rock as huge bodies slam into the Moon. I give some more detail on the origins of these great mountain-rimmed basins later.

The fact that the full Moon is bright from limb to limb, rather than becoming duller toward the edges as might be expected from a sphere that was reflecting sunlight, has always been commented upon. It is due to the fact that the surface is broken up into tiny fragments by the impact of meteorites. Thus, it is covered with a myriad of reflecting surfaces, like a giant bicycle reflector in the sky. This may explain how the ancient Greeks arrived at their concept that the heavenly bodies were made of shining crystal.

A thin veneer (typically a few kilometres thick) of basaltic lavas (the maria) form the familiar dark features of the 'Man in the Moon'. Figure 30 shows these lavas that have filled the circular craters punched into the white highland crust. Lavas rise to the surface of the Moon and the planets because the liquids are less dense than the surrounding solid rock. Lavas on the Moon are more common on the near side. There they can more easily reach the surface because the crust is thin. In contrast, they are rare on the far side of the Moon because they mostly fail to reach the surface through the much thicker crust.

The controversy over the nature of these dark patches on the face of the Moon is interesting. Before the Apollo missions, they were often thought to be sediments, dust, or asphalt-like material because they are so dark and flat-lying. Only a few acute early observers identified them as basaltic lavas, as they have few signs of eruptive centres or anything looking like a terrestrial volcano. This lack of familiar terrestrial volcanoes is a consequence of the low viscosity of the lunar lavas. This enables them to flow as easily as oil for hundreds of kilometres on slopes of only a degree or two. This was unexpected. Basalt lavas

Figure 30. The contrast between the lunar highlands and the lunar maria
is well shown in this view of Mare Ingenii on the lunar farside. The rugged
white regions are the old crust of the Moon, mostly made of feldspar.
Large circular craters have been punched into the crust by impacts. Later
these holes have been flooded with lavas, producing the smooth grey plains
of the maria. These have a few small craters on their surface from later
impacts. The large circular crater, filled with mare basalt, is Thomson, 112
kilometres diameter, in the northeast sector of Mare Ingenii, 370 kilometres
diameter. The crater in the right foreground is Zelinsky, 54 kilometres
diameter, excavated in the old highland crust. The sequence of events that
have produced this scene from oldest to youngest, is (1) formation of white
feldspar-rich highland crust (2) excavation of Ingenii basin (3) formation
of Thomson crater (4) formation of Zelinsky crater (5) flooding of Ingenii
basin and Thomson crater with basaltic lava and (6) production of small
impact craters on the smooth mare surface, including a possible chain of
secondary craters (NASA AS15–87–11724).

on the Earth are viscous, more like toffee than engine oil. The Moon
contained many other surprises for geologists and geochemists.
These taught us that every planet is different. This has freed us from
the constraining influence of ideas based on our local geology.

Inside the Moon

The basaltic lavas were erupted from deep within the Moon and tell us that the interior is composed of zones of variable composition. These zones formed during the solidification of the Moon as various minerals precipitated from the melted rock. There is growing evidence for the existence of a small metallic core, but it may form only a few per cent of the volume of the Moon. Among the most surprising results from the Apollo samples was the demonstration of ancient magnetic fields, which are now dead. This topic has caused more controversy than any other, despite strong competition from other lunar problems. The maximum strength of the ancient lunar field was probably about half that of the present terrestrial magnetic field. The most likely possibility is that the field was generated internally during the freezing of a fluid iron core.

The composition of the Moon

Study of the first Apollo sample returned from the smooth basaltic plains of Mare Tranquillitatis found some unusual chemistry compared with rocks on the Earth. The interior of the Moon is bone-dry without a drop of water. The possible presence of ice in deeply shadowed craters at the south pole, reported by the Clementine Mission radar, might reflect a recent addition to the lunar surface of water from comets, but the observation is very uncertain. The Moon is heavily depleted in the most volatile elements, for example lead and chlorine. It is this feature of lunar geochemistry that makes the Moon unique among satellites and planets. The Moon has lost the moderately volatile elements such as sodium to a lesser degree, and is enriched in the refractory elements such as calcium and aluminum.

Life on the Moon

Although it is true that the Moon has no atmosphere as we understand the term, there is an extremely thin atmosphere, consisting of only a sprinkling of atoms, which are sputtered off the surface by radiation from the Sun. Of more interest is the question of possible life on the Moon. The following comments come from old sources, but they scarcely require any updating. Thus, Christian Huygens (1629–1695) said 300 years ago that 'the Moon has no air or atmosphere surrounding it as we have. I cannot imagine how any plants or animals whose whole nourishment comes from liquid bodies, can thrive in a dry, waterless, parched soil'.[4] James Breen (1826–1866) made a further contribution 150 years ago when he noted that 'In the want of water and air, the question as to whether this body is inhabited is no longer equivocal. Its surface resolves itself into a sterile and inhospitable waste, where the lichen which flourishes amidst the frosts and snows of Lapland would quickly wither and die, and no animal with a drop of blood in its veins could exist'.[5]

However, our technical civilisation is capable of providing a comfortable environment there. Residing on the Moon, shielded by a few metres of the lunar soil, which is moved as readily as beach sand on Earth, would not be very different from living in Antarctica, except that oxygen and water would be needed. The polar regions seem the most likely site for an inhabited lunar base. Some sites at the poles on crater rims enjoy continuous sunlight. There one could have a permanent solar power supply, a consideration that will over-ride all other requirements in selecting a site. Other regions would have to endure two weeks of darkness each month.

Evolution of the Moon

The broad features of both lunar composition and lunar evolution are well understood. They are now known much better than for the Earth. The views about the Moon before the *Apollo* missions cast their revealing light had led to the view that the moon was a primitive object. This was mainly because of its low density relative to the Earth. The surprising thing was that the samples returned from the surface of the Moon turned out to be very highly enriched in many elements. They look more like samples from the surface of the Earth than what was expected in a primitive body. Indeed, the highland crust was so enriched in refractory elements that models appeared of crustal formation invoking the late plastering on of a layer, like adding icing to a cake.

Melting of much of the Moon, and development of the thick lunar highland crust, took place immediately after the formation of the satellite. This vast mass of molten rock has been termed the 'magma ocean' and a highly energetic and rapid mode of origin for the Moon is required to account for it. The crystallisation of this ocean of melted rock is understood in principle. Feldspar was an early mineral to crystallise. It floated, due to the low density of the feldspar crystals and the bone-dry nature of the silicate melt, and formed a thick feldspathic crust. 'Rockbergs' of feldspar may have swept together like icebergs, so accounting for the differences in crustal thickness between the near and far sides. A tiny lunar core of iron formed in the centre. This took in what little nickel, platinum and gold had reached the Moon. The lunar rocky mantle was fully crystallised within a few million years. Zones formed of different minerals, from which the lavas that darken the face of the Moon were derived much later. The final dregs resulting from the crystallisation of the Moon were highly enriched in those elements that could not fit into the common minerals. This material, referred to as KREEP, was mixed into the crust by impacts and accounts for the very high concentrations of elements such as potassium and uranium in the crust of the Moon. However, in the absence of the recycling that accompanies plate tectonics, this material was not concentrated into ore deposits that we could mine.

Lunar craters

A major insight from the study of the Moon was the evidence for massive impact cratering early in the history of the solar system. One of the most striking features of the lunar surface is the evidence of meteorite impacts at all scales, from large basins, ranging from hundreds to over a thousand kilometres in diameter, with concentric rings of mountains, down to tiny micron-sized pits caused by micrometeorites hitting grains lying on the surface. Although the larger craters were long thought to result from volcanic activity, their origin due to the impact of asteroids, comets and meteorites, was established beyond doubt about 30 years ago, following much controversy over impact versus volcanic shaping of the face of the Moon. Laplace, with his usual insight, noted in 1796 that 'all is solid at the surface of the Moon, on which some have thought they perceived the effects and even the explosion of volcanoes'.[6] This piece of wisdom went unheeded by many, who kept reporting events on the Moon right up to the first *Apollo* landing in 1969. Such observations, made at the limits of resolution of telescopes, remind one of the martian canals of Lowell.

In contrast to the Earth, the major structural features of the lunar crust were formed by great impacts. This resulted in the common occurrence of circular landforms. These range from the mountain rings to the circular basins filled with dark basaltic lava. Following lunar formation and crustal solidification at about 4440 million years ago, a population of large objects (up to a hundred kilometres in diameter) struck the Moon over the next 500 million years. These impacts formed at least 80 ringed basins with diameters greater than 300 kilometres on the solid crust. Smaller colliding bodies created an additional 10 000 craters in the range 30–300 kilometres in diameter, and there is a host of lesser sized craters. This bombardment declined steeply in intensity after the formation of the final ringed basins (Imbrium and Orientale) at about 3850 million years ago.

Similar heavily cratered surfaces on Mercury, Mars and the satellites of the giant planets show that these catastrophic events were not unique to the Moon. The impact of all these bodies at high velocities resulted in a zone of fractured and brecciated rubble on the Moon perhaps a few kilometres thick on the highlands. Meteorite impacts have occurred at a much slower rate on the Moon since the termination of the great bombardment. The youngest such major event on the Moon was the formation, about 100 million years ago, of the crater Tycho, 85 kilometres in diameter. Tycho formed due to the impact of a small mountain-sized body, a few kilometres in diameter. Material ejected during this impact forms the bright rays, which extend across the visible face, and are such a spectacular feature of the full Moon, particularly when viewed though binoculars. This large impact on the Moon is not far removed in time from the similar catastrophe on the Earth that wiped out the dinosaurs and much else.

There is a great contrast between the long-term survival of craters on the Moon and their rapid destruction on the Earth. About 35 million years ago, an asteroid struck at what is now the mouth of Chesapeake Bay on the eastern seaboard of the United States. It dug a crater 90 kilometres in diameter, about the size of Tycho. The collision sprayed out some glassy fragments, which are now found as tektites as far away as Texas. In contrast to craters on the Moon, which remain untouched until destroyed by later impacts, the young crater in Chesapeake Bay was rapidly filled by sediment and covered from view. It has only recently been discovered by a combination of geophysical probing of the subsurface structure and by study of samples obtained by drilling. Meanwhile, Tycho, which is about three times older, sits untouched, with its beautiful rays radiating across the face of the Moon.

The origin of the Moon

The problem of the origin and evolution of the Moon turns out to have been an interesting intellectual exercise. Attempts to explain complex natural phenomena call for a wide variety of skills and the Moon presented a particularly difficult example. Preconceived notions and prejudices frequently overrode both data and common sense.

Following the sample return, pre-*Apollo* theories for the origin of the Moon all failed for various reasons. Hypotheses in which the Earth captures an already formed Moon were abandoned. It turns out to be very difficult to capture the Moon into its present orbit around the Earth. In such a model, the curious chemistry of the Moon had to form somewhere far away. Putting problems out of sight does not solve them.

The similarity in density between the Moon and the Earth's silicate mantle has fuelled the speculation, dating back to George Darwin (1845–1912), the fifth child of Charles, that the Moon was formed from the Earth's rocky mantle following core formation on the Earth.

Such 'fission hypotheses' that derive the material for the Moon from the Earth's mantle encounter two basic difficulties. The spin of the Earth–Moon system, although large, is insufficient by a factor of about four to allow fission to occur. A second objection is more telling. Fission models have become the most readily testable of all following the lunar sample return, as they predict that the chemistry of the Moon should bear some recognisable signature of the rocky mantle of the Earth. However, the composition of the Moon stubbornly refuses to fit, despite heroic attempts by geochemists to match them.

Double planet models that form the Moon and Earth in association possess the twin difficulties of failing to account for the high spin of the Earth–Moon system, and of readily accounting for the density difference. They do, however, account for the similarity in oxygen isotopes between Earth and Moon.

Yet another model formed the Moon from a ring of rocky debris produced by break-up of incoming asteroids as they come within the Roche Limit. This process is supposed to result in a ring of broken-up rock debris around the Earth. Their tougher iron cores stuck together and crashed into the Earth. This sounds more like a process for making rings rather than satellites. Both these hypotheses should be general features of planetary formation and lead to the presence of moon-like satellites everywhere, so that Venus should be graced with a satellite.

None of these theories accounts for the unique nature of the Earth–Moon system, for the strange lunar orbit and for the high spin of the Earth–Moon system. Like ships hitting an uncharted rock, they all sank from this defect. Uncommon objects like our Moon may require an uncommon origin.

A large impact on the Earth

The assembly of all the information about the Moon into a model for the origin of the Moon to satisfy all the conditions has proven difficult, but something approaching a consensus is now being reached. The high spin rate of the Earth and Moon cannot arise through small impacts. However, one large impact could account for it. However, one had to leap several orders of magnitude beyond the scale of the great impact scar of Mare Orientale, to propose the single large impact hypothesis for lunar origin (see Figure 31).

Figure 31. The current model for the origin of the Moon. Computer simulation of the formation of the Moon by the oblique collision with the Earth of a body 0.14 Earth masses at a velocity of 5 kilometres/s. Both the Earth and the impactor have differentiated into a metallic core and a silicate mantle. Times following the impact is shown in the boxes. Following the collision, the impactor is spread out in space, but the debris clumps together through gravitational attraction. The iron core of the impactor separates from the silicate mantle and accretes to the Earth

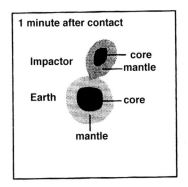

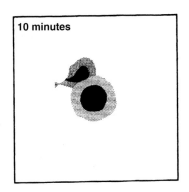

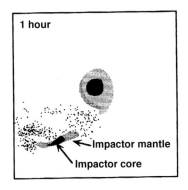

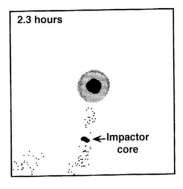

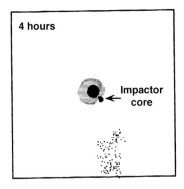

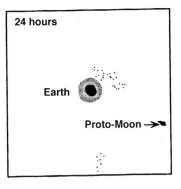

about four hours after the initial encounter. About 24 hours following the impact, a silicate lump of about the mass of the Moon is in orbit around the Earth. This material comes mainly from the rocky mantle of the impactor. (Courtesy A. G. W. Cameron and W. Benz, Smithson. Astrophys. Inst.)

This concept that required an overview of many different fields. The impactor had to be bigger than Mars. When this theory was first proposed, this was too large a lump for most to swallow. However, this single impact theory resolves many of the problems associated with the origin of the Moon. The best conditions are for the impactor to have a mass about 15 per cent of that of the Earth and for it to hit the Earth at a velocity of five kilometres a second. This body, as well as the Earth, is assumed to have formed a metallic core and rocky mantle by the time they collide. The collision breaks up the impactor and much of the rocky mantle goes into orbit about the Earth. The metal core of the impactor separates from the rocky mantle. This rocky material is accelerated and the metal core decelerated relative to the Earth. The metallic core falls into the Earth within about four hours. Whether the remaining rocky material forms several moonlets, or coalesces rapidly into the Moon is not known, but the end result is that the Moon is totally melted.

Only a small amount of material from the Earth's mantle eventually ends up in the Moon. Since the hypothesis derived the Moon principally from the rocky mantle of the Mars-sized impacting body, this notion cut the Gordian Knot[7] tying the Moon to the Earth. The impact event was sufficiently energetic to vapourise much of the material that went to make up the Moon. This naturally explains such unique geochemical features as the bone-dry nature of the Moon and the extreme depletion of very volatile elements. A final consequence of the giant impact model for lunar origin is that the event was energetic enough to melt the mantle of the Earth. Such melting appears to be an inevitable consequence of the accretion of large planets.

Of course a prime requirement for this useful hypothesis is that there was a supply of bodies of the right size to hit the Earth. Fortunately for the model, there is plenty of evidence from the tilts of the planets for the previous existence of Mars-sized bodies in the inner solar system. Unique events are difficult to accommodate in most scientific disciplines. However, although the details of the Moon-forming event are not predictable, massive collisions early in solar system were common. One just happened to have the appropriate mass and velocity and hit the Earth at the right angle to provide us with the Moon, which has been such a source of inspiration to poets and princesses alike.

The effect of the Moon on the Earth

The Moon-forming and other massive collisions had several effects significant in making the Earth a suitable abode for life. Any early dense atmosphere was removed by the impact. The tilt of the planet, another probable consequence of the collision, produces the seasonal variations. The rather rapid rotation of the Earth, in contrast to the slow rotation of Venus, provides a day–night temperature and light variation less stressful to the development of life than much longer or shorter periods might be. The day has become a little longer over time as the Moon has receded from the Earth, so slowing the rotation of our planet by tidal forces as it retreated. About one billion years ago, when the Moon was about 35 000 kilometres closer, the day was about 18 hours long.

The Moon may also stabilise the tilt of the Earth, in contrast to that of Mars, which wobbles about. Widely varying tilts would cause variations in the amount of sunlight that regions would receive. The associated changes in climate, would presumably produce a stressful environment for life. This adds another factor to consider in the search for habitable planets in other planetary systems. The Moon also plays a unique role by raising substantial tides in the Earth's oceans.

Mercury

A red herring

It is often thought that tiny Mercury is difficult to see with the naked eye. Indeed it is close to the Sun and so is mostly lost in the glare. However, it is often visible away from city lights, as a morning or evening star. Of course it does not rival Venus in brightness, being further from us and only one twentieth of the mass of the Earth.

The strange nature of Mercury has only recently been appreciated. Like the Moon, it has played the role of a red herring in our attempts to understand the solar system. The high density of Mercury, the highest for any planet, was a particular trap. The planet is closest to the Sun of all the members of the solar system. So it fits in

Figure 32. The internal structure of Mercury, showing its immense core of iron and its thin rocky mantle. The surface is covered with craters caused by the impacts of meteorites, asteroids and comets.

very well with a grand overall scheme. In this scenario, the planets near the Sun have been cooked and lost gas, water and other volatiles. Further out, where it is cooler, the planets are less dense, and have a lot of gas. While there are some elements of truth in this simple and appealing picture, there are many difficulties, as I have pointed out earlier.

Too much iron, too little rock

The combination of high density, magnetic field and rocky surface tell us that this little planet has melted and separated into a large iron core and a rocky mantle (see Figure 32). The basic problem is that Mercury is so dense that it must contain about twice as much iron relative to rock as the other inner planets.

We know that the surface is rocky and covered with a layer of dust and rubble that looks very like the surface layer on the Moon. Photographs from the space mission that ventured there in 1974 are often mistaken for pictures of the Moon. More sophisticated measurements of the light reflected from the surface look much like what is seen on the white feldspar-rich highlands of the Moon. The surface is very old, covered with craters and impact basins. It is as at least as old as the highland crust of the Moon, which is well over four billion years old. Thus, the iron that is responsible for the high density is not on the surface, but must reside in a large metal core. This core is enormous compared with the rest of the planet. It occupies about three quarters of the planetary volume. In contrast, the core of the Earth, although much larger than that of Mercury, occupies only about one quarter of the volume of the interior of the Earth.

An unexpected magnetic field

A totally unexpected discovery was that Mercury has a magnetic field. It is very much weaker than the field for the Earth, but presents us with some interesting dilemmas. The magnetic field of the Earth is thought to form as a sort of dynamo in a liquid metal core. But Mercury is so small that the general opinion is that the planet should have frozen solid aeons ago. Curiously, Venus, which is nearly the size of the Earth, has no magnetic field. Perhaps the core of Mercury contains enough sulphur to keep it molten, but then why does Venus not behave in the same way? Magnetism is almost as much of a puzzle now as it was when William Gilbert (1544–1603) wrote his classic text

'Concerning Magnetism, Magnetic Bodies and the Great Magnet, Earth' in 1600.

An unusual orbit

There are some other unique features about Mercury. The orbit is the most eccentric of any planet. It might have been supposed for a small planet closest of all to the massive Sun, that its orbit would be the most regular in the solar system. However, the orbit of the tiny planet as it circles round the Sun is tilted at over 7 degrees to the Earth–Sun plane. Both these facts need to be explained in any theory for the origin of Mercury.

Being so close to the Sun, the spin of Mercury has become locked in by its giant neighbour, just as the Moon always shows us the same face. Mercury's case is a bit different to that of the Moon. It rotates three times for every two orbits around the Sun, ensuring that its surface gets to be baked at temperatures hot enough to melt lead. Only the floors of some deep craters at the poles are shielded from the glare. As there is no effective atmosphere, these deep holes remain below the freezing point of water. Earth-based radar reflections may indicate some trapped ice, which could have arrived via comets. It seems strange that ice might exist so near the Sun. Like the report of ice on the Moon, we need much more confirmation before setting out, like travellers in the Sahara, to find what may be a mythical oasis.

The barren plains of Mercury

Although the surface is mostly pockmarked with craters, there are some extensive plains on Mercury. The origin of these plains is much disputed. Some think that they formed as sheets of rock debris thrown out from impacts. Others propose that they are plains made of lava like the dark patches visible on the face of the Moon. After all, since the other inner planets produce basalt lava in abundance, why should Mercury be an exception? However, the plains of Mercury are too bright to be made, like those on the Moon, of dark basaltic lava.

In view of the probable violent early history of Mercury, it seems unwise to use data derived from other bodies to interpret its surface composition. Current models for the origin of Mercury suggest that much of its rocky mantle was removed by a giant impact that disrupted the planet. Thus, the composition of the reassembled debris might be different to that of the rocky mantles of the other terrestrial planets. Perhaps the mantle of Mercury is more refractory than that of the Moon, Venus, Mars or the Earth. Accordingly, any lavas that were erupted from it could indeed be unique.

Earlier on, there was a similar argument over similar light-coloured plains on the highlands of the Moon. Many thought that they were due to volcanic eruptions. However, when visited by the *Apollo 16* astronauts, they were discovered to be debris sheets from large impacts. This experience of trying to interpret the surface composition of the Moon from photographs should make us cautious. Before we went to the Moon, one worker at least realised that 'a surface cannot be characterised by its portrait. The Moon remained inscrutable at all scales'.[8]

My own opinion is that the smooth plains of Mercury are formed from sheets of debris ejected by giant collisions. Some support for this has been obtained from spectra in the microwave region from the surface of Mercury. These spectra show no evidence of iron or titanium, two key elements present in basaltic lavas. The spectra greatly resemble those from the lunar highlands. So possibly Mercury has a crust of feldspar like the Moon.

A shrunken planet

One of the most interesting features about Mercury is the evidence that it had undergone a slight contraction in its radius at a very early stage of its history. A unique feature of the planet, not seen on other planets, is the presence of large fault scarps that cut across the surface. They are about one kilometre in height and several hundred kilometres long. They tell us that the planet has shrunk by somewhere between two and four kilometres in radius. Much of the contraction is due to the cooling and solidification of the mantle and crust around the large iron core. The scarps cut the older craters and the intercrater plains. Younger, fresher craters cut the scarps and some of the smooth plains appear to be younger. This shrinking thus occurred towards the end of the massive bombardment and must have happened before 4000 million years ago. Following this initial contraction of the planet, the radius of Mercury has been unchanged for at least four billion years.

A collisional origin for Mercury

The driving force behind previous proposals to account for the strange features of Mercury has been the need to account for the high density of the planet. This unique high density of Mercury has had a fatal attraction for solar system modellers and inventors of grand unified theories for the origin of the solar system. The density of Mercury was another trap just like the similarity between the density of the Moon and that of the rocky mantle of the Earth. That had provided a tantalising and ironic clue to lunar origin, and misled workers from George Darwin, in the last century, to the present day.

Those making early attempts to explain the composition of Mercury, and in particular its high iron abundance, were seduced by its proximity to the Sun. This was because the planet apparently anchors one end of a sequence extending from the high-density inner planets out to the low-density outer solar system bodies. This variation appeared to be consistent with a decrease in temperature outwards from the Sun. However, if the origin of the high density of Mercury is due to some cosmic accident, it is no longer relevant to grand models for the origin of the solar system.

The most likely model to account for the strange composition of tiny Mercury is that much of the rocky mantle was lost during a massive collision. Current estimates are that the impacting body was about one fifth of the mass of Mercury, hitting the planet head-on at 20 kilometres per second. In this scenario, the initial mass of Mercury would be about twice its present value. The rocky material would be smashed into pieces less than a centimetre in size, and mostly swept away to finish up in the Sun, Venus or the Earth. We may have some bits that were originally in Mercury. The tough metal core hung together and wrapped itself, like a cosmic beggar, in a thin cloak of rock. This model also provides for the strange orbit of Mercury. A bigger collision might have left us with only the iron core and so an iron planet at which to wonder. Thus, Mercury is another battered survivor from the turbulent early history of the solar system. Many similar-sized bodies must have existed in our region before the final sweep-up, with most of them finishing up in Venus or the Earth. However, Mercury with its high density is so strange that it falls into a special category, unique even by the standards of the solar system. This tiny planet is a good example of the refusal of most of the planets and satellites to be put into neat pigeonholes.

6
Causes and effects

The solar system is not very tidy and the arrangements of the planets are not what might be expected from a simple condensation from the original disk of gas and dust. Neither, for that matter, does it look like the work of a competent designer with omnipotent powers. What is the reason for this diversity? Why is the Earth is the only possible abode for life in the solar system? This planet is exceptional; clearly Venus or Jupiter wouldn't be agreeable to us. How did the Earth achieve this status? This raises other questions. Here, I discuss whether the solar system is unique. What do other planetary systems look like? Finally, how will it all end?

The collisions of asteroids and comets with planets

An untidy system

The planets all circle the Sun in the same sense, a consequence of the original rotation of the disk of gas and dust from which the system formed. One might expect that if the planets formed from such a rotating disk, that they would all be upright. They should either spin at the same speed, or in some tidy mathematical sequence, just like the regular spacing of Bode's Rule. Although the planets mostly spin on their axes in the same sense, anticlockwise when viewed from above the north pole, there are exceptions to this regular arrangement. Venus rotates slowly backwards, while Uranus lies on its side. Moreover, the planets are all tilted, and rotate at different speeds. All bear some signature of having experienced unique events. Why is this so?

The essentially random nature of impacts has changed our philosophical outlook on the origin of the solar system. If large collisions are a characteristic feature of the final stages of the accumulation of the planets, then it is not possible to predict the details of the events. Such collisions occurred at all times and stages in the history of the solar system. This process began with the sticking together of grains in the primitive disk of dust and gas. It continued with the growth of bodies that eventually reached the size of small planets. Innumerable impacts occurred during the sweepup of these smaller bodies into the planets. These culminated in the massive final collisions that tilted the planets and started them spinning at different rates. The largest impacts occur toward the end of the process. Some collisions spun out disks from the giant planets from which satellites formed. I discussed earlier how a giant impact produced the Moon and how another stripped the rocky mantle off Mercury.

The tilts and spins of the planets

The different tilts and spins of the planets constitute the best evidence for massive impacts in the early solar system. No model involving condensation from a disk in an orderly manner can account for the rather untidy situation in which the planets now find themselves. It would be extraordinary if all planets were tilted to the same degree, just as much as if all planets were identical. If the planets showed zero tilts or some obvious regularity in their spins, it would be possible to entertain an orderly origin for them.

The tilt of the Earth provides us with the seasons because of the variable amount of sunlight received over different areas. At present Mars has a similar tilt to that of the Earth, but over time the red planet wobbles through perhaps 60 degrees. Venus, in great contrast, has only a very small tilt. Opinions differ about the cause. Was Venus hit head on and stopped in its tracks? Perhaps it never suffered a giant impact, for its slow backward rotation could be the result accumulation from a collection of small bodies. The slow rotation of Venus may indeed be the primitive state for most planets, and their varying spins, like their tilts, are all due to massive collisions late during their formation.

The tilts of Jupiter and Saturn may be due to a combination of collisions and perhaps warps in the gaseous nebula. Jupiter has only a slight tilt, but Saturn is inclined at nearly 30 degrees to the common plane of the solar system, more so even than the Earth. Uranus and Neptune have significant tilts. The most extreme case is Uranus. This planet is 14 times more massive than the Earth. It takes the impact of something the size of the Earth to knock a planet of this size over. Uranus has a collection of nine rings and 15 satellites that all rotate around its equator. These must all have formed after the planet rolled over. Thus, the planets bear, not only on their battered faces, but also in their varying tilts and spins, silent witness to the trauma surrounding their birth.

Curiously enough, the plane in which the planets lie is tilted at seven degrees to the equator of the Sun. This is rarely discussed. Perhaps some late torque twisted the gaseous nebula away from the plane of the Sun's equator and is responsible for part of the tilts of the giant planets.

Both Mercury and the Moon represent special cases in the inner solar system. Both owe their unique character to the effect of large collisions. Like Mercury, but for a different reason, the Moon is an anomaly in the solar system. It has a very low density, in contrast to the high density of Mercury. One body has too little iron and the other has too much. Thus, collisions can produce strange bedfellows.

Pluto and its very large satellite Charon are not only in an eccentric and highly inclined orbit, but rotate around each other, at right angles to the rest of the system. As we saw earlier, such a curious situation, which seems difficult to produce in an orderly system, is a likely result of a massive collision.

A universal bombardment

How do we know about about the former existence of large bodies? These have now vanished. The evidence comes from the observation that all of the older surfaces on planets and satellites are saturated with craters. The lunar surface, visible through the smallest telescope or binoculars, is the classic example. Spacecraft photographs show that from Mercury out to the satellites of Uranus a massive bombardment struck planets and satellites. Craters of all sizes are present. The range from micron-sized pits due to impact of tiny grains on lunar samples, up to giant ringed basins the size of France or Texas (see, for example, Figure 11). The extent of this early bombardment on the Moon is revealed by the presence of at least 80 big basins with diameters greater than 300 kilometres. Another 10000 craters are in the size range from 30–300 kilometres. These formed before the main bombardment ceased about 3850 million years ago. Since a similar barrage struck the Earth, this accounts for the absence of rocks older than that age on our planet.

The history of the debate over whether volcanoes or meteorites caused the craters on the Moon is another fascinating chapter in science. It is full of misconceptions, misidentifications and faulty conclusions. Unfortunately, I must leave most of this fascinating topic to the historians of science. Because we live on a planet on which erosion removes craters rather quickly, the significance of impacts in solar system history has been only slowly appreciated. Even now, there are pockets of resistance to the idea among the more conservative geologists.

In earlier times, it was thought that the craters on the Moon were mostly due to volcanoes. This view was held by many right up to the manned landings of the *Apollo* spacecraft in 1969. Among the many puzzles that the craters presented to students in previous times, the fact that the craters on the Moon were mostly circular was one of the most difficult to explain. The reasoning ran that meteorites would hit the Moon at all angles, and so should produce mostly oval craters. It was not until this century, with its unfortunately close acquaintance with high explosives, that the cause of the circular nature of craters was understood. An early instructive example comes from the Great War of 1914–1918 (which became, with hindsight, the First World War). A mine was exploded on July 1, 1916 at La Boiselle in a futile attempt to breach the German lines on the Somme. The crater was 85 metres wide and 25 metres deep, complete with a five metre high rim and an ejecta blanket of chalk. It closely resembled a crater due to meteorite impact. There are many more recent examples.

Cosmic impacts are more destructive than our puny efforts. Thus a rocky fragment 250 metres in diameter, around the size of a football stadium, hitting the Earth at perhaps twenty kilometres a second, has the explosive energy of a thousand megatons of TNT. No matter at what angle it hits the Earth, it will bury itself and explode like a bomb.

The resulting explosion digs a circular crater five kilometres in diameter and about one kilometre deep. Within a few minutes, the surrounding countryside for several kilometres from the crater will be covered with a deep blanket of rubble and broken-up rock thrown out of the crater.

About every 20 or 30 million years an asteroid in an earth-crossing orbit hits the Earth and forms a crater twenty kilometres across. The impact of a large Apollo or Aten asteroid with the Earth would produce a major catastrophe. It could wipe out life on Earth. If life survived, evolution might go off in some other direction. Perhaps bizarre forms like those in the Burgess Shale, which I talk about later, might appear again as evolution starts tinkering in a new direction. The extinction of the dinosaurs and much else 65 million years ago at the end of the Cretaceous Period was due to the impact of an asteroid or comet about 10 or 15 kilometres in diameter. That devastating collision produced a crater in Mexico over 200 kilometres in diameter.

The early intense bombardment

The most spectacular landforms discovered by spacecraft are probably the giant impact basins such as Orientale on the Moon. However, like the immense martian landscapes of Valles Marineris and Olympus Mons, which are so impressive when photographed from orbit, they would be less striking on the ground. Early workers, notably G. K. Gilbert (1843–1918), Ralph Baldwin (b. 1912) and Harold Urey, all drew attention to the circular nature of the Imbrium basin on the Moon. It's the size of Texas. They considered that it formed when a large asteroid hit the Moon. The later discovery of the Orientale basin provided a nearly perfect example of a ringed impact basin. With five concentric rings of mountains, it is like a great bullseye, 900 kilometres across. Photos of Mars and Mercury show similar immense basins surrounded by rings of mountains. Radar mapping of Venus showed that even that thick cloud-shrouded atmosphere provides no protection if the impacting bodies are large enough.

We can estimate how frequent these great early impacts were. The best evidence comes from the Moon, where we know the age of the surface from the samples that came back from the *Apollo* missions. However, perhaps the Moon is a special case. As it slowly retreated from the Earth, did the Moon run into a ring of bodies in orbit around the Earth? There is still a lot of debate among workers in these fields. Some insist that there are also differences in cratering rates between the inner and outer solar system. Others believe that the craters we see are only the tail-end of a bombardment that was uniform throughout the solar system.

Great big basins

Small craters, a few kilometres in diameter, have the shape of a simple bowl. When the craters are a little larger, they have a curious feature. There is a mountain peak in the centre of the basin. These central peaks were often mistaken for volcanoes. These mountains formed as the floor of the crater rebounded following the smashing impact and explosion of the meteorite. There are good examples of such structures, now mostly heavily eroded, on the Earth. Here, we can measure how far the strata have rebounded upwards. A well known example is Gosses Bluff, in central Australia, where the flat-lying rocks formerly deep beneath the surface have been bounced upwards nearly four kilometres, where they form an impressive peak of jumbled up strata four and a half kilometres across. Erosion has removed most of the outer rim of the crater, which originally was over twenty-four kilometres in diameter, leaving only the central peak to bear silent witness to the ancient catastrophe.

Bigger impacts produce even more distinct effects. Next to appear in this increasing scale of catastrophic events are the so-called peak-ring basins. These are distinguished by a ring of hills on the floor of the crater that is perhaps 100 kilometres across. This ring has formed when the central peak has collapsed and spread out like a giant ripple. All these catastrophic events happen within a few minutes following the impact.

Then there are stupendous collisions that form very large basins that are hundreds of kilometres in diameter. They are surrounded by impressive rings of mountains. On the Moon, these were given names such as the Apennines, the Carpathians and the Jura Mountains after their terrestrial namesakes. The origin of these puzzling circular mountain ranges, although clear to astute observers such as Gilbert, Baldwin and Urey, was finally made obvious to all just before the *Apollo* missions. This was due to the superb photographs of the Orientale basin obtained by *Lunar Orbiter IV* in May 1967 (see Figure 11).

Many models have been advanced to account for these large mountain-rimmed basins. As one worker remarked, 'the origin of multiple rings, both internal and external, has nearly as many interpretations as there are investigators'.[1] Most likely, the large basins were formed by collapse into a deep central cavity that the asteroid dug initially. The rings of mountains then formed as the sides of the cavity collapsed around concentric faults.

Two examples on the Earth of such large impact structures are the Sudbury basin in Canada, famous as a source of nickel, and the Vredefort structure in South Africa. In the latter example, the continental crust, which we regard as so solid, has been overturned to a depth of twenty kilometres. Although some skepticism has remained among geologists, decisive evidence has recently confirmed the origin of these immense structures by the impact of meteorites, asteroids or comets. The Earth cannot produce such amounts of explosive energy in an instant. The eruptions of the largest volcanoes are puny by comparison. By now these ideas are generally accepted and no one has seriously questioned the idea that the large crater in Yucatan, Mexico, which is the scar remaining from the dinosaur-killing event, was due to an impact.

Addition and subtraction of atmospheres and oceans

Our models for building the Earth do not make many predictions about the nature of the primitive atmosphere of the Earth. Of course the Earth formed after the gas in the original disk had been driven away. A collection of dry rocky bodies was left from which to make the Earth. The atmospheres of the inner planets were added later, probably from comets. Thus, our atmosphere is entirely secondary. Accordingly, it is not surprising that the atmospheres of the inner planets are so different from our ideas about the composition of the primitive dust and gas from which the solar system formed. The most likely candidate to have retained a primitive atmosphere anywhere in the solar system is Titan, the large satellite of Saturn.

However, another major factor enters to complicate the problem.

Probably the growing planets suffered several giant impacts and repeatedly lost any atmospheres that they had evolved or collected. Losing atmospheres is clearly easier on small planets. It is an illustration of the law that the poor get poorer. Clearly, such processes would only be effective before the close of the period of heavy bombardment about 3850 million years ago. Since that time, the atmospheres of Venus and the Earth have been stable against such major loss.

If all of the atmospheres of the inner planets are secondary, and have been subject to random removal by impacts, then they won't contain much information about the primitive disk of gas and dust. Instead, the evolution of each planetary atmosphere is unique. Thus, while the present atmospheres of the planets are interesting as scientific problems in their own right, they will probably never provide insights into the origin of the individual planets. This is one of the casualties of our improved understanding of the effects of large impacts.

All the evidence from the meteorites is that the primitive mineral components were dry. Water was driven out of the inner solar system and condensed as ice far away from the Sun, at a 'snow line' in the vicinity of Jupiter. Thus, the terrestrial planets seem to have accumulated from dry rubble. Probably the only way for water to get to the inner planets is from comets coming from the outer reaches of the solar system.

On average, comets hit the inner planets at higher velocities than asteroids or meteorites from the inner solar system. These impacts are so energetic that they are just as likely to remove an atmosphere as to deposit one. Thus, it seems plausible that on a bigger body such as the Earth, much of the water may have been delivered by comets. All these calculations depend heavily on the assumed rate of impacts of comets. This in its turn is beset with uncertainty.

Earth and Moon: the continuing bombardment

Compared with Venus, Earth has had many more smaller craters form due to its much thinner atmosphere that allows smaller meteorites to reach the surface of the planet. Most craters here are removed quickly on geological time scales by erosion. Because of this, only about 150 craters, stretching back in age over the past two billion years, have been found on our planet. The eroded remnants of many more must await discovery, because the number of impacts on the Earth must be close to that observed on Venus. On that basis, we can calculate that over 10 000 bodies capable of forming craters over five kilometres in diameter must have hit our continents or the oceans over the past four billion years of Earth history.

The objects which now hit the Earth are either comets coming from the outer reaches of the Solar System, or asteroids and meteorites. These are perturbed from their orbits in the asteroid belt by collisions or gravitational effects of Jupiter. Occasional meteorites, knocked off from Mars or the Moon by large impacts, are also swept up by the Earth. Bodies less than about ten metres in diameter are burnt up in the atmosphere. Due to their high speed, they have energies equivalent to a few megatons of high explosive such as TNT.

The cause of the event over Tunguska in Siberia on June 30, 1908 has been the subject of vast speculation. An explosion occurred at 7.30 am local time at an altitude of five kilometres and released energy equivalent to about 20 megatons of TNT. The resulting shock wave blew down the Siberian forest for over 1000 square kilometres. Although many bizarre explanations, including the arrival of some anti-matter, have been proposed to account for this sudden explosion, the reality is more prosaic. The cause is now known to be a stony meteorite colliding with the Earth. The meteorite was only the size of a small building, about 60 metres in diameter, but it was travelling at perhaps 20 kilometres a second. If the meteorite had exploded over New York, it would have completely demolished the city. Such an event happens somewhere on the Earth perhaps every 300 years. Fortunately, the statistical chances of such a large meteorite hitting a great city occur only once in a million years.

As I pointed out earlier, a local catastrophe would be caused by a body 250 metres diameter impacting at 20 kilometres a second and digging a crater five kilometres across. The energy released would be equivalent to 1000 megatons of TNT. Such an event is likely to happen about every 10 or 20 thousand years. Catastrophes affecting the whole world are likely to be caused by collisions with objects between about one and five kilometres in diameter.

How much material reaches the Earth at present? Estimates vary widely, but the generally accepted figure is that about a hundred tons of meteoritic material, fortunately mainly as dust, falls on the Earth every day. This sounds like a lot if it was all delivered to your backyard, but it adds only a trivial amount to the mass of the Earth over geological time.

The extinction of the dinosaurs

Charles Darwin (1809–1882) commented that 'no fact in the long history of the world is so startling as the wide and repeated extermination of its inhabitants'.[2] The extinction of many species at the boundary between the Cretaceous and Tertiary strata, 65 million years ago, has been often remarked upon. The successful dinosaurs, who had held sway for 160 million years (see Figure 33), vanished along with a great multitude of species.

What is particularly striking is that the extinction occurred at a geological instant. Where the geological strata are well preserved, as in drill cores from oceanic sites, the wipeout of species occurs at a knife-edge. No slow change in the atmosphere or oceans can produce such a sudden catastrophe. This disaster left most evolutionary niches open to be exploited by the survivors. The mammals emerged and took over as the dominant land-dwelling species, leading to the eventual development of *Homo sapiens*.

The scientific consensus is that the ultimate cause was the impact of an asteroid between 10 and 15 kilometres in diameter. Something the size of Mt Everest travelling at at least 20 kilometres a second hit the Earth. If it was a comet, it could have been going twice as fast. The evidence for this impact at the end of the Cretaceous Period is particularly strong. The most striking and first discovered fingerprint was a world-wide spike of the rare element iridium in the thin layer of clay that marks the end of the Cretaceous in the geological record. Iridium is rare in the Earth's crust but common in meteorites. Next to be found in the clay, again over most of the world, were grains of quartz and feldspar that showed fractures due to a massive shock. It was as though they had undergone a giant hammer blow. Then a lot of melted material that we now find as glassy fragments had been sprayed around. Another team found soot from wildfires ignited by the event. There was so much soot that most of the forests on Earth must have burnt to supply it. All these observations strongly support

Figure 33. About 130 million years ago, *Brontosaurus*, one of the the largest of the dinosaurs, contemplates a comet. *Brontosaurus* died out in the late Jurassic Period, 60 million years before the great extinction of his reptilian relatives and much else at the end of the Cretaceous Period.

an impact event. No geological process, such as a great outpouring of lava, is capable of accounting for these facts.

The centre of the impact is near Chicxulub on the Yucatan Peninsula in Mexico. The crater, or rather ringed basin, formed by the explosion was at least 200 kilometres in diameter. Over the next 50 million years, it was slowly buried under the limestone that was deposited in the warm Caribbean Sea. The asteroid struck a particularly deadly location. It dug in to several kilometres of strata containing carbonates and sulphates. These had formed by evaporation in a bay of the ocean that had intermittently dried up and been reflooded with sea water. The result was that hundreds of billion of tons of sulphur dioxide and carbon dioxide were blasted into the atmosphere. The sulphur dioxide combined with water vapour to form sulphuric acid. The resulting acid rain killed all near surface marine life that had shells of carbonate. These included the tiny foraminifera that lived in their countless billions. Their shells were easily dissolved by the acid. Nearly all the species vanished in a geological instant. Even those species, such as diatoms, with shells of more resistant silica, did not survive, probably on account of the 'nuclear winter' effect. Several months of darkness were caused by the combination of dust and smoke in the atmosphere. First it was too cold and then it became too hot. Following the 'nuclear winter', temperatures rose due to the large amount of carbon dioxide injected into the atmosphere from the carbonates in the target area, contributing further environmental stresses. These probably disrupted the entire food chain and so were also responsible for the extinction of the giant marine reptiles, such as the icthyosaurs and plesiosaurs whose graceful forms we admire in museums.

It is interesting to speculate how various mythologies might account for the removal of such a dominant and successful race as the dinosaurs. The ancient Greeks could readily have credited Zeus with hurling a thunderbolt to remove the dinosaurs. The Old Testament Jehovah, who drove Adam and Eve from the Garden of Eden, and barred their reentry with a flaming sword, might have been equally dissatisfied with his reptilian creation. Given omnipotent powers, it would not be too difficult to direct the asteroid to wipe the slate clean, and so give the mammals their opportunity. However, a few survivors would have to be left, such as the serpent needed to tempt Eve. *Homo sapiens* seems equally capable of such draconian measures. It is for such reasons that proposals to prepare atomic weapons to deflect possible incoming asteroids should be ignored. History gives every expectation that these weapons, if not used for warfare, would be employed to arrange for the arrival of an asteroid on top of one's enemy. One is reminded of Hitler, who in the last days of the Third Reich in 1945, sought to destroy the German people as unworthy of his vast purpose, in a *Götterdämmerung*.

A closer call: the end of the trilobites (see figure 34)

This catastrophe at the end of the Permian Period, 250 million years ago, dwarfs the extinction at the Cretaceous–Tertiary boundary. It came close to extinguishing life on Earth. Seventy per cent of vertebrate families on land and 90 per cent of species in the oceans became extinct. It is possible that there was a double extinction, with two events that were about five million years apart. The first seems to have

Figure 34. *Dalmanites*, a trilobite about 5 cm long, from the Silurian Period. The trilobites, of which many thousands of species are known from the fossil record, flourished for nearly 300 million years, before being wiped out in the great extinction at the end of the Permian Period, 250 million years ago.

killed about three quarters of marine species. The later one, right at the end of the Permian Period (and of the Paleozoic Era) finished off over 80 per cent of those species that were left, or had evolved in the meantime.

The causes appear more complex ('a tangled web'[3] according to one investigator) than the Cretaceous impact. Apart from impact, as yet unproven, possible causes involved changes in the arrangement of the continents at that time. This could have led to unstable climates and drying out of marine basins. Coupled with massive volcanic eruptions and an increase in carbon dioxide in the atmosphere, these events may have led to a catastrophic collapse of life.

All this is much less clear in the geological record than the episode that removed the dinosaurs. For that event, we can study complete sections recovered from drill cores in the deep oceans. Strata at the Permian–Triassic boundary are less extensive than those that mark the end of the Cretaceous. The sea floor of the earlier Permian times has all vanished back into the mantle, carrying with it much of the evidence. There are of course many other massive extinctions throughout the geological record. In addition to these catastrophic events, evolution takes its toll, discarding species so rapidly that they have long formed useful markers for geological strata.

Extinctions at regular intervals?

Considerable interest was aroused by the suggestion that mass extinctions in the fossil record showed a regular pattern. They appeared to occur about every 26 million years. Such regularities are rare in nature. What was the cause? For a while, it was generally ascribed to a series of cometary impacts at regular intervals. These were thought to be due to periodic showers of comets coming from the Oort Cloud. These showers were believed to be triggered by the influence of a large undiscovered outer planet (our old friend Planet X) or by an undiscovered dark companion of the Sun. This mythical creature came to be called the 'Nemesis' star, as it seemed to be doing deadly damage to life.

There is considerable doubt both about the existence of this star and of the periodic nature of the extinctions in the geological record. The statistical evidence for periodicity is very weak and the patterns of extinction cannot be separated from that due to chance. It seems that incorrect statistical treatments have been used by those who relate extinctions to impact events occurring at regular intervals. This was confirmed by Richard Grieve (b. 1943), a Canadian geologist, who has made the most detailed study of the terrestrial cratering record. Scott Tremaine (b. 1950), a Canadian astrophysicist, notes that

> there has been a long and dismal history in astronomy of spurious periodicities which have been claimed in many types of data. Hidden periodicities used to be discovered as easily as witches in medieval times, but even strong faith must be fortified by a statistical test. The most likely situation is that the alleged periodicity in the extinction record is spurious, and that any relationship between impacts and extinctions likely involves single, large, random impacts of comets or asteroids.[4]

Thus, the supposed correlation of extinctions with impact craters and the 26 million year periodicity of crater-forming impacts turns out to have arisen through faulty statistics. The patterns of mass extinction appear to have arisen by chance and cannot be distinguished from the effects of a random process.

Life and the anthropic principle

The Earth as an abode for life

The orbit of the Earth, which every child now learns is elliptical, is of course rather close to circular. This convenient arrangement would be expected in an orderly system. If the orbit were highly elliptical, life would have to adapt to extreme temperature differences as the Earth approached to and receded from the Sun. But many of the conditions that are so favourable to the existence of life were due to accidental events. Among the many random events that occurred during the formation of the solar system, a massive body larger than Mars crashed into the Earth. The rocky mantle of the impactor was splashed out into orbit around the earth. It formed the Moon. This dramatic event played a significant role in making this planet a suitable abode for life. Any dense early atmosphere was removed by the impact, allowing our present agreeable atmosphere to form.

The presence of the Moon seems also to have stabilised the tilt of the Earth. This has given us relatively stable climates. The tilt of the Earth gave us the seasons, spring, summer, autumn and winter, which are themselves a stimulus to evolution. These changes give us philosophical concepts of cycles and renewal as the amount of sunlight changes from winter to summer. Another consequence of the great collisions is the rather rapid rotation period of the Earth. This contrasts with the slow rotation of Venus. Much longer or shorter days and nights might be more stressful to the development of life. The Moon also plays a unique role by raising substantial tides in the Earth's oceans. The effects are particularly dramatic along shorelines in estuaries and narrow bays, and provide a rich environment for life to develop, as a glance into any tidal pool demonstrates. The chance formation of the Moon provided this useful environment, for, without our satellite, tides raised by the Sun would be minute.

The features that make the Earth uniquely suitable for the development of life are perhaps most dramatically demonstrated by comparing the Earth with our twin planet, Venus. In contrast to our oceans, lakes and rivers, there is only a trace of water in the atmosphere of Venus. While this atmosphere consists mainly of carbon dioxide, on the Earth carbon dioxide is mostly locked up in limestones.

On the surface of the Earth, the growth of great continental land masses allows for the development of mountains, broad plains, extensive forests, sweeping savannas and major rivers. Large short-lived lakes appear following episodes of continental glaciation. Continental drift provides many changing patterns, such as shifting climatic zones, formation and destruction of mountain ranges, and extensive shallow seas as continental shelves are flooded. All these produce a multitude of stimulating environments for land-based life. If the continental crust had not provided the setting for the development of this diversity, evolution, restricted to small islands, would have taken a different course. Birds might have become the dominant land-dwellers, as in New Zealand or Mauritius before the arrival of humans.

The principal difference in the surface conditions on Earth and Venus appears to be related to the presence of abundant water on the Earth's surface. The operation of plate tectonics, continental drift and the formation of the continents themselves is mostly due to the presence of water. The Earth is thus a very dynamic planet. On the dry surface of Venus, as on Mars, Mercury and the Moon, plate tectonics does not occur, and barren basaltic plains, pocked with craters like a Great War battlefield, are the common landscape. So the surfaces of the other planets constitute a No-Man's-Land of planetary proportions, which is as equally unfriendly to life as was the Western Front.

The Earth is about the right distance from the Sun to make this an agreeable and habitable planet. This question is often referred as the Goldilocks problem after the girl in the fable who tasted the porridge of the three bears. Venus is too hot, Mars is too cold, but the Earth, like Baby Bear's porridge, is just right. This, however, is a bit simplistic, since much more than distance is involved. The surface temperature of the Earth that we find so agreeable, is maintained by a 'greenhouse' effect, which traps heat. Without water and carbon dioxide in the atmosphere, the surface temperature would average 18 degrees below zero Centigrade, and the world would resemble Siberia in the depths of winter.

Venus is too hot mainly because of the 'greenhouse' effect of its thick atmosphere of carbon dioxide, not because it is so much closer to the Sun. Were it not for the greenhouse effect that traps the heat, the surface of that roasted planet would be below freezing. Its clouds reflect so much of the energy coming from the Sun that Venus absorbs only a little more solar radiation than Mars. Even the thin atmosphere of Mars adds a few degrees to the surface temperature of that frozen desert.

The width of the zone around the Sun in which a habitable planet can reside in our solar system is quite narrow. Various estimates range from about a tenth of an AU to about half an AU around the orbit of the Earth. But making a habitable planet depends on a complex set of factors, of which distance from the Sun is only one. The amount and composition of the atmosphere and the nature of the cloud cover are critical. So it's not just a matter of getting an earth-sized planet at the right distance from a star. A host of other factors are involved.

However, that is not all. The Earth has maintained a rather even climate for four billion years. Indeed, the Earth seems if anything to have been a bit warmer three billion years ago. At this point, the astronomers present us with a problem. The Sun at that ancient epoch had been shining for only a short time. Theory tells us that it would have produced about one quarter to a third less light than it does now. This is the famous 'faint early Sun' problem. The astronomical theory seems robust enough, firmly based on the physics of nuclear fusion of hydrogen to helium. One might therefore expect that the early Earth would be a frozen waste, which warmed up slowly through the ages as the Sun increased its output.

In contrast, the geological evidence is quite definite that running water, eroding the surface and producing water-laid sediments complete with ripple marks, evidence of tides and much else, has been present throughout these vast epochs. The fascinating occasional climatic fluctuations that we call ice ages have occurred only at rare intervals throughout the geological record. Mostly the climate of the Earth has been much milder than that of the present unstable interglacial period that began about 10000 years ago as the ice sheets retreated.

Various explanations have been offered to explain how the Earth managed to maintain an even climate despite the faint sunshine that the astronomers insist upon. Usually some kind of greenhouse effect is invoked. Methane, ammonia and carbon dioxide have all been suggested. However, an early atmosphere rich in the carbon dioxide that is mostly called for, can have the reverse effect. Clouds of dry ice are likely to form in the high atmosphere, reflecting the Sun's rays, so leading to a 'snowball' Earth that might become a permanent condition. Clearly, some delicate balance of conditions allowed for the benign conditions that allowed life to arise and flourish.

Thus, the Earth evolved through a series of chance events into an abode suitable for the origin and evolution of life. Other planetary systems are unlikely to contain clones of the Earth, and must be expected to differ substantially from our own. Chance events continued to affect the evolution of life throughout geological time, the most dramatic being the massive collision with an asteroid 65 million years ago, which wiped out the ammonites, the graceful plesiosaurs, the great dinosaurs and much else at the Cretaceous–Tertiary boundary.

This great disaster cleared the way for the evolution of mammals to fill the vacant spaces. It is sobering to realise that, but for this cosmic accident, *Homo sapiens* would not have evolved at all, and this account would not have been written.

The origin of life

The first essential for life as we know it is the presence of carbon and related elements. These chemical elements, however, were not there 'at the beginning'. Only hydrogen (which constitutes nearly three quarters of the universe), helium (about a quarter) and a trace of lithium were produced in the Big Bang. The heavier elements, including carbon, oxygen, nitrogen, phosphorous, iron and other elements essential for life and also for making planets, were formed later by nuclear reactions inside stars and in supernova explosions. The nuclear furnaces in the stars have formed them on a scale that would have astounded the medieval alchemists. As a star comes to the end of its life, it sheds mass or explodes. The newly formed elements are dispersed out into the gas and dust of interstellar space. They provide the material from which new stars are born.

Clearly life could not begin until elements such as carbon, potassium and phosphorus became relatively abundant. It took several billion years and many generations of stars to produce enough of these elements for life to begin. By the time that the solar system formed, the heavier elements, including the life-forming elements, formed about two per cent of the gas and dust in the solar nebula. Thus, life is not some all-pervasive presence in the universe, but just another set of chemical compounds that had to await the appropriate conditions. Precursor molecules, such as amino acids and nucleic acids, are common enough that given the appropriate conditions on a planet, life might originate anywhere in the universe. Thus, it seems likely that life may arise anywhere given the right mix of chemistry and environment and so be common in the universe. The development of intelligence is another matter that I discuss later.

This view is often referred to as 'reductionist' and conflicts with the idea that life has a vital, mystical or ethereal component, outside of physical reality, that is commonly referred to as the 'soul', an idea that goes back at least as far as Plato. The concept of the soul as part of the mind, or the 'ghost in the machine', seems to be a relatively recent concept, mostly due to the great French (or Dutch, for he spent his most productive years in Holland) philosopher, Rene Descartes.

Francis Crick (b. 1916) in his entertaining book *The Astonishing Hypothesis* presents a different view. It is that 'you' are, in fact, no more than the end result of the behaviour 'of a vast number of nerve cells and their associated molecules'.[5] However, this seems to me, in the words of Thomas Jefferson (1743–1836), 'to be self-evident' rather than 'astonishing', and I continue this discussion on a thoroughly reductionist theme on the basis of scientifically observable and testable models.

Estimates for the date of the origin of life on Earth converge mostly around 4000 million years ago, but the subject is beset with uncertainty.[6] The famous experiments of Stanley Miller (b. 1930) and Harold Urey in the 1950s produced abundant amino acids, possible precursors of life. This feat was achieved by electrical sparks, simulating lightning, in a mixture of hydrogen, ammonia and methane that was thought to mimic an early terrestrial atmosphere containing those gases. However, there is considerable geological evidence that by about four billion years ago the early atmosphere consisted of water vapour, carbon dioxide and nitrogen. If our atmosphere in fact arose by degassing or recycling from volcanoes, then the early atmosphere was unlikely to have contained large quantities of hydrogen, ammonia and methane. Ancient volcanic rocks are not very different from modern lavas and probably gave off water and carbon dioxide like present volcanoes. Other models suppose that life arose in Darwin's 'warm little pond'. In such an environment, the complex organic compounds produced in the atmosphere by lightning, solar UV radiation, in addition to those brought in by meteorites, might achieve the requisite jump into reproducing species. Clay minerals are often thought to provide a sort of template.

Two problems have combined to make these scenarios less attractive. In those remote epochs, there is a scarcity of continental crust and only a few islands provide dry land or tidal pools. More serious is the devastating rain of asteroids and comets. Judging from the numbers and ages of craters on the Moon, impacts of massive bodies capable of forming basins hundreds or thousands of kilometres in diameter continued for several hundred million years. Such globally sterilising events were common before 4000 million years ago. The development of life must have been frustrated by the continuing bombardment which smashed up any early crust that had tried to form on the Earth, while the largest collisions removed any early atmosphere.

Because of these problems with the development of life on the surface, attention has now become focussed on how life could have sought shelter from the raining bombardment. Deep under the oceans, life might arise in relative safety and colonise the surface regions only when the bombardment slackened. In this case, conditions for the origin of life may have existed as early as 4300 million years ago, shortly after the Moon-forming impact. Volcanic vents, often referred to as 'black smokers', are common deep under the oceans at the mid-ocean ridges. They are a rich source of strange life-forms. Among the most primitive organisms are sulphur-bearing bacteria, and experiments have demonstrated that complex organic molecules can be formed under these conditions.[7] These environments were widespread early in Archean times, when most lavas erupted under the oceans in the absence of appreciable land masses.

Like gloves, complex organic compounds possess both left- and right-handed forms in equal proportions. If organic molecules had continued to arrive from outer space, both left and right-handed forms would have been delivered from meteorites. However, life on Earth has been curiously selective. The amino acids in living organisms are only 'left-handed'. In contrast, the nucleic acids are all 'right-handed' molecules. Why is this so? No one knows, but it looks as though a single event or perhaps selective process was involved. So the origin of life looks like a one-shot affair. Perhaps there was some obscure evolutionary advantage in selecting the forms, or maybe it was another chance event that persisted once the pattern was set. One is reminded of the 'qwerty' arrangement of computer keyboards. This originated because of the need to separate frequently used letters to avoid jamming problems in early mechanical typewriters. The problem has long disappeared, but we are left with the fossil order, despite the existence of more efficient arrangements for keyboards.

Fred Hoyle (b. 1915) and his coworkers have interpreted some features in the spectra of Comet Halley as evidence for the presence of bacteria. Such an identification would have profound implications for the origin of life, since such bacteria could arrive 'ready-made' on the Earth from outer space. However, the fatal flaw in this entertaining concept is that the identification is non-unique. Many organic molecules show features that provide an excellent match to the spectral fingerprints of bacteria. It is not difficult to find look-alikes for the spectra of bacteria among the several million organic compounds that are known.

However life arose, there is some evidence that it had become established on Earth by 3850 million years ago. Ancient sedimentary rocks in Greenland that formed in that remote time bear a debatable signature of possible microbial life in the ratios of their carbon isotopes. The presence of sedimentary rocks at that distant epoch also tells us that water was present on the surface of the Earth. What is surprising is that life not only appeared so early, but that it developed so rapidly. Life seems to have been clearly in existence a few hundred million years later. Stromatolites (bacterial fossils) are preserved as fossils in the Warrawoona Group, 3450 million years old, in Western Australia. Curiously, similar stromatolites are forming at present not far away in Shark Bay, on the west coast of Australia, an outstanding tribute to the ability of life to survive on this planet for over three and a half billion years. This also tells us that conditions on the surface of the Earth have not drastically changed in the region over that immense length of time.

So life seems to have been well adapted from its beginning. Biology has built an astonishing variety of organisms (some three million current species of beetles) using relatively simple modules, just as many organic chemical compounds and complex minerals have been constructed from a few basic building blocks.

On the development of intelligent life

In this book, I have argued that the likelihood of developing copies of this planet, or of our solar system, are remote. However, it is of interest to observe the progress of evolution on this one example that we have available to us and to enquire whether something resembling *Homo sapiens* might arise elsewhere. This question is often listed as being of major philosophical importance.

The first major observation is that evolution has no necessary direction or plan to produce us as its ultimate achievement. For the first two billion years, life on this planet was restricted to simple bacteria and archea. The emergence of organisms with structurally complex cells, the so-called eukaryotes, occurred for little understood reasons about 1800 million years ago. As Preston Cloud (1912–1991) commented 'the appearance of the eukaryotic cell was a Proterozoic triumph – the main event of biological evolution after the origin of life itself'.[8] Vast numbers of multicellular organisms arose from this chance twist. Species mostly have a relatively short life on geological time scales, varying from 100000 years to four million years. This leads to estimates that somewhere between 10 and 50 billion species have existed on the Earth. The current population of perhaps 10 to 30 million species displays a marvellous diversity in adapting to a myriad of environments. Some species of course live for much longer times and have survived effectively unchanged for immense periods of time. The modern brachiopod genus *Lingula* is a classic example. It is similar in appearance and structure to its ancestor, *Lingulella*, which lived over 500 million years ago in the Cambrian Period. Sharks, which currently number about 300 species, first arose in the Devonian Period. Having successfully occupied their ecological niche, they retained their basic shape through a succession of different species for the next 375 million years.

However, despite this great flowering of life, intelligence seems to have developed only among the vertebrates, and there, rarely. Among the 24 orders of mammals, high intelligence seems to have arisen in only one, in primates. Why is this so? Clearly high intelligence has little evolutionary advantage, for it has appeared once in tens of billion attempts. As Ernst Mayr (b. 1905), the biologist, has pointed out,[9] even the development of high intelligence may not lead to the ability to communicate with distant planets. Only one of the 20 or so civilisations that have arisen on Earth in the past 5000 years has developed the technology to communicate with other possible life forms elsewhere.

But even on this well-endowed planet, there was nothing preordained about the emergence of *Homo sapiens* on the plains of Africa. Three separate continents were available on the Earth on which the later stages of the evolution of land animals could evolve.[10] All these vast areas shared the benign conditions on this planet that make it such a comfortable environment for life. When life first invaded the land in the late Silurian and Devonian Periods about 400 million years ago, the scattered continents were slowly uniting into a single land mass, which we call Pangaea. During the next few hundred million years, as plants and animals evolved and the dinosaurs became dominant, this great mass began to split up. A large southern continent, called Gondwana (after an historic region of central India), sailed away. This in turn slowly fragmented into familiar pieces that now appear on our maps of the world. Australia, Antarctica and South America, carrying their cargo of animals and plants, broke away leaving Africa in isolation. Australia separated from the frozen southern continent and departed northward. Africa and India also travelled north at a rate of a few centimetres a year, finally ramming into Europe and Asia and creating from this titanic collision the mighty mountain chains of the Alps and Himalayas.

So were formed the three continental masses on which the later evolution of land animals proceeded independently. Australia, isolated from the rest of the world, produced the weird marsupial animals that puzzled early explorers. In South America, the land animals and their fossil ancestors intrigued Charles Darwin by their differences from his familiar European species, but the South American monkeys, primates like us, never left the trees. Africa, in addition to the splendid array of lions, antelopes, zebras, giraffes and the rest that we all admire, managed to produce another unique species, *Homo sapiens*. On the other continents no species remotely resembling us arose. The sobering conclusion is that even when everything else in the environment was perfect, blind chance still ruled the development of intelligent life.

When the remote possibilities of developing a habitable planet are added to the chances of developing both high intelligence and a technically advanced civilisation, the odds of finding 'little green men' elsewhere in the universe decline to zero.

Life on Mars?

Mars is fascinating to us since it is the only other site in the solar system that has some approach to earth-like conditions. A major reason for sending the NASA *Viking* Landers to Mars in 1976 was to attempt to discover whether life was present on that planet. Unfortunately, unanticipated conditions on the surface defeated this attempt to make exobiology a science that contained some subject matter. The *Viking* landers carried three experiments based on biology. All gave ambiguous results. Thus, substantial amounts of oxygen were evolved when water was added to the soil samples. These strange results have been generally attributed to the presence of a strongly oxidising component in the soil. These experiments illustrate the extreme difficulty in designing tests to identify life outside the Earth.

A secondary lesson from the *Viking* Landers is that one should not make the experiments on board spacecraft too complicated. The experience from the biological experiments on the *Viking* Landers is that if you did not get the expected answer, you received data that you could not interpret. Examination of a few grams of the martian soil, returned to Earth, would have quickly told us what was causing the strange reactions in the biological experiments.

In fact, a decisive experiment on the *Viking* Landers was carried out by a mass spectrometer designed to identify organic compounds. Nothing was found down to parts per billion levels. In their total absence, no terrestrial-type life forms can be expected. This contrasts with the surface of the Moon, on which even a few parts per million of organic compounds were present. This may not sound very much, but it is over 1000 times greater than in the soil of Mars. As such molecules must also be added to Mars by meteoritic or cometary infall, this is an apparent paradox. There must be an efficient way by which organic molecules are being destroyed on the surface of Mars. Probably any such organic material is destroyed by the presence of oxidising chemical compounds in the soil.

Following the negative results from the *Viking* Landers in 1976[11] the question of life on Mars was summarised by a cartoon: 'It has been a disappointing year! Nothing on TV, nothing in Loch Ness and nothing on Mars'. Nevertheless, raw hope remains and the search is now focussed on looking for traces of fossil life around possible early hot springs. This is fuelled by the discovery of bacterial life in extremely hot and cold environments on Earth. These illustrate the ability of life to adapt to nearly any environment on this planet.

This search has now been given an extra impetus. Some possible faint traces of early bacterial life have been claimed to have been discovered in a meteorite from Mars. This rock was excavated from beneath the surface by a meteorite impact 16 million years ago, and eventually fell to Earth in Antarctica 13 000 years ago. It was picked up from its icy bed in 1984. The rock itself is very old, four and a half billion years old, dating effectively from the beginning of the solar system. Curiously, it is far older than any of the other meteorites from Mars. Tiny tube-like forms are observed on globules of carbonate in veins of minerals. These formed within fractures that resulted from meteorite impacts, and the veins in which the tiny bacteria-like forms are found are perhaps a billion years younger than the time when the original rock formed.

Several lines of evidence have been claimed in support of the notion of primitive life in this meteorite. These included the presence of oily organic compounds (polycyclic aromatic hydrocarbons or PAH) and various mineral associations, all of which, taken together, might indicate the presence of some ancient biological activity. Chains of evidence are notorious for weak links and the mineral evidence can be explained by inorganic processes, while PAHs are widespread contaminants in the Antarctic ice. Neither seem capable of constituting the 'smoking gun'. Perhaps the most difficult problem is that the size of the tiny tube-like 'organisms' is about the size of our smallest viruses. There is room to accommodate only a few hundred atoms. Terrestrial bacteria are about a thousand times larger than these enigmatic martian structures. Our experience on Earth is that much larger molecules are apparently needed for self-replicating organisms.

Considerable skepticism is warranted. In the 1960s, there was much excitement over the discovery of 'organised elements' in primitive meteorites. These looked like pollen grains, which in fact they were. They had come from the field in which the meteorite had fallen, or perhaps had drifted into the museum case in Paris where the meteorite had lain for over a hundred years.

However, if the presence of ancient life on Mars is eventually confirmed, it will have considerable philosophical interest. It will tell us that life may arise anywhere if the chemical conditions are right. It is of course a very long way from simple bacteria to the 'little green men' that so many people want to believe in. Furthermore, if life did arise on Mars, and died out, then life was clearly unable to influence the environment on Mars and so adapt the planet to ensure survival. This would constitute both a test of and a disproof of the Gaia hypothesis of James Lovelock that the Earth is alive.

Was the universe designed for us?

One would scarcely have arranged the haphazard series of hazardous events that have happened in the solar system, if your ultimate objective was the emergence of *Homo sapiens*. Indeed, if the universe were designed to produce our species, it is an unbelievably inefficient mechanism. Thus, it took over ten billion years to produce as a final result of this grand design, such interesting specimens of *Homo sapiens* as Pol Pot, Genghis Kahn, Attila the Hun and Adolf Hitler. There are innumerable lesser examples of appalling behaviour by our species. Probably this is a heritage from hunter–gatherer or more primitive, perhaps Neanderthal societies. The results are visible most nights on television, an astonishing invention that demonstrates our technical brilliance. However, this marvellous medium of communication has become mostly a wasteland of trivia and violence. This tells us in turn of the intellectual limitations of *Homo sapiens* and of the inability of our species to adapt quickly enough to changing conditions and to adopt the civilised behaviour patterns now required for the survival of our highly developed technical society.

It was William Paley (1743–1805) in his book, *Natural Theology* (1802), who made the clearest argument that there must be a designer, using the famous analogy of someone finding a watch lying on Hampstead Heath (a natural area of 800 acres, now surrounded by modern London). The watch could hardly have arisen by chance and immediately suggested the presence of a someone who had designed it. Paley elaborated his arguments by carefully considering the details of the human body. His most famous example was the eye. How could it have arisen by chance?

Charles Darwin commented that

> the old argument of design in nature, as given by Paley, which formerly seemed to me so conclusive, fails, now that the law of natural selection has been discovered. We can no longer argue

that, for instance, the beautiful hinge of a bivalve shell must have been made by an intelligent being, like the hinge of a door by a carpenter. There seems to be no more design in the variability of organic beings and in the action of natural selection than in the course which the wind blows[12]

All the studies of biology since Darwin have confirmed this view. Thus, before Darwin it was impossible to answer the arguments that everything was designed, but, as Richard Dawkins (b. 1941) comments 'Darwin made it possible to be an intellectually fulfilled atheist'. He comments further that

> Natural selection, has no purpose in mind. It has no mind and no mind's eye. It does not plan for the future. It has no vision, no foresight, no sight at all. If it can be said to play the role of watchmaker in nature, it is the *blind* watchmaker.[13]

The biologists have opted for evolution and the operation of chance events. They have abandoned the search for a designer, having no need of that hypothesis, ever since Darwin showed how complex organisms could arise. Now it is chiefly among the physical scientists that support for a designer universe resides. This is usually stated as various forms of the anthropic principle, which I discuss a little later. This situation is ironic, since one would think at first glance that physics, with its precise mathematical formulae, is much simpler than the bewildering variety that one encounters in biology. Thus, it is interesting that most of those who have supported the search for extra-terrestrial intelligence have been physicists. The biologists, acutely aware of the random chances of evolution, have in general been skeptical.

Stephen Jay Gould (b. 1941) has pointed out that another quiet revolution has occurred in our thinking about evolution. This has come about largely from a reexamination of the Burgess Shale, in British Columbia.[14] This formation, which was originally a muddy sea floor in Middle Cambrian time, 530 million years ago, has preserved a fauna of soft-bodied animals. Such preservation is exceedingly rare in the geological record, for obvious reasons. Mostly, only the hard parts survive the many perils, such as being eaten, before becoming preserved as fossils. The preservation of the entire community now found in the Burgess Shale seems to have been due to a geological accident. A submarine landslide swept over the sea floor burying all these strange creatures. Only a few examples of such complete preservation are known in the entire geological record. A famous example is the first bird, *Archaeopteryx*, complete with preserved feathers, which was found in Germany in the Solnhofen Limestone, formerly a Jurassic coral lagoon.

Apart from containing many well known fossils, such as trilobites, what is interesting about the Burgess Shale is that it contains animals belonging to no known phylum. Phyla are of course the major divisions of the Animal Kingdom and there are currently somewhere between 20 and 32 of them, depending upon which biologist you talk to. The Burgess Shale contains another dozen or so organisms that are so distinct that they deserve to be classified as distinct phyla (depending upon which palaeontologist you talk to). These bizarre forms (to us) never appear again in the geological record. They have no modern counterpart. The grotesque animals that we can inspect, for example in the extensive collections in the National Museum of Natural History in Washington, DC, represent failed evolutionary experiments. Thus, the present millions of species represent only one set of possible life forms. Evolution could had wandered off in another direction with results impossible to predict. Even so, the diversity of those that we now observe, is bewildering. The past forms were more so. We regard elephants with their long and useful trunks as remarkable and perhaps unlikely beasts. However, our two types of elephants (Indian and African) are the only survivors of 300 species of the formerly extensive family of the *Proboscidea*, which included the giant woolly mammoths from Siberia and the various dwarf elephants of Crete, Cyprus and Malta.

Clearly survival is random and but for accidents, evolution could have taken a totally different direction. As Richard Dawkins has remarked, 'a true watchmaker has foresight'[15] but evolution is makeshift, adopting any expedient solution that serves for the moment. Certainly when contemplating a newly born baby of our species, one is struck by the inefficiency of its digestive system. Surely a designer could have come up with a better solution.

Dawkins[15] notes that eyes of bewildering complexity have developed independently at least 40 and perhaps up to 60 times during the long course of evolutionary development. Estimates for the time taken to evolve a lens eye require perhaps only half a million years, a mere blink of geological time. But Dawkins notes that our eyes are a second-best compromise. The receptors in our eyes, and in all vertebrates, are like photocells (all 166 million of them) with their connecting wires coming out on the front. A sensible design would connect them to the brain on the side away from the light. However, they work well enough. The changes needed to reverse the wiring carry no immediate advantage, and might result in degrading our vision while the changeover was being accomplished. Thus, although our remote descendants would benefit, evolution has no interest except in the here and now. Curiously enough, the octopus has the wiring the right way round. Its superior wiring does not prevent it from being eaten by our super-successful species.

The anthropic principle

Homo sapiens has been relegated to a distant corner of the universe. It was bad enough when Nicolaus Copernicus put the Sun, rather than the Earth, at the centre of the world. Worse was to follow. Early last century, geologists discovered the unfathomable 'dark backward and abysm of time',[16] of an extent unimaginable on human lifespans. They were followed by Charles Darwin who placed *Homo sapiens* among the animals. Then Edwin Hubble (1889–1953), who found that even our Milky Way galaxy, whose edge-on view is at least visible to us, was only one speck amongst a great host. The space telescope that bears his name has now enlarged even this stupendous space by another couple of orders of magnitude.

This new knowledge from science seems to have damaged our egos perhaps deeper than we realise. Surely after all, we were meant to be the Lords of Creation, occupying a central place in the cosmos, rather than the inhabitants of some obscure, even if interesting backyard. However, the human ego is stronger even than that of the household cat, *Felis domesticus*. Now self-interest has attempted to bring our existence back to a place of central importance in the universe. Amongst the more respectable attempts is the *anthropic principle*,[17] and it finds comfort for us among the constants of physics. It is the latest version of design arguments that go back to Aristotle.

There are variations of the anthropic principle, usually referred to, like tea or alcoholic drinks, as strong or weak. Like theological dogma, these principles are a bit difficult to pin down with any precision. The strong variety seems to say that the universe must have the properties to enable life to arise, and so life must arise at some stage in the universe. This would imply a conscious act by a creator, and so be scientifically untestable. It should be noted, however, that the principle does not specify that the life which arises as a consequence must be intelligent, although no doubt this was intended by the authors of the idea. This oversight recalls the fate of Tithonus, the mortal lover of Eos, Goddess of the Dawn in Greek mythology. She persuaded Zeus to make him immortal, but forgot to ask the god to grant him eternal youth. Eventually Tithonus became helpless with old age, but he talked incessantly and so was shut away, perhaps a cautionary tale for philosophers.

The weak variety of the anthropic principle is a little more flexible and says that the physical properties of the universe have taken on those values to enable human beings to exist and measure them. This merely states that we live in a universe that we can observe, and seems to be a statement of the obvious.

The list of physical properties that make our existence possible is certainly impressive. One example is the curious chemistry of water and the density of ice. Most solids are denser than their liquid forms, but ice floats on water. If it didn't, it would sink to the bottom of lakes and oceans, and never thaw, making life as we understand it very difficult.

The masses of protons and neutrons are just about right to enable hydrogen to form. Then there is the famous example of the difficulty of forming the element carbon during element synthesis in stars. The existence of a metastable state in the binding energy of the carbon nucleus enables it to exist just long enough, not only to preserve the element carbon, but to enable the chain to proceed to make heavier elements, such as oxygen. Without that quirk of nuclear physics, we would not be here. There are many others. Of course, one might ask that if the purpose of a creator is to produce *Homo sapiens*, why not alter this delicate condition, or make carbon and the rest of the elements in the beginning in some variant of the Big Bang. And why take so long, for that matter?

Next there are the interesting mathematical connections among various atomic constants, usually referred to as the large numbers coincidence. Why do ratios involving 10 to the 40th power keep appearing? Is the large numbers coincidence just a 'brute fact' with no more significance than the fact that the Sun and the Moon have almost exactly the same apparent size in the sky? Is that some kind of cosmic joke, put there to fool us, or is it merely a remarkable coincidence? Maybe it is just the way things are, a 'brute fact' or 'blunt truth', unrelated to our existence.

Possibly most of the fundamental constants were fixed as a consequence of the Big Bang, just as the value of pi (π) gives the relationship between the diameter and circumference of a circle, or the stability of the chemical elements is fixed by the ratio of protons to neutrons in the atomic nucleus. If the constants were different, or if no carbon had been formed, some other form of intelligence might have arisen, just as the Periodic Table would contain some other interesting elements if the binding energies of protons and neutrons were different. It is of course true that all these properties are essential to our existence, just as the correct value of π is essential to the construction of an automobile. Curiously a value of 3.0 for this constant appears in the description of the Temple of Solomon in the Old Testament, a work commonly considered inerrant.[18] Certainly it would be difficult to construct a workable wheel on the basis that π is 3.0, rather than 3.14159.

It is ironic that we now see apparent design in physics, just as William Paley saw the hand of the designer in biology 200 years ago. Darwin explained the reasons for biology. Are the physicists still waiting for their Darwin to explain the reasons behind the apparent evidence of a designer who set up the physical constants for us? It seems possible that eventually the 'large numbers' coincidence, the ratio of photons to baryons, and so forth will be able to be calculated from some 'Grand Unified Theory of Everything'. Then these curious coincidences will not need to be accounted for by the anthropic principle, any more than we now need the god Thor to account for thunderstorms.

The anthropic principle suffers from that fatal defect of scientific hypotheses of being untestable. For this reason, it is a philosophical curiosity somewhat like the Gaia Hypothesis that states that the

Earth is alive. It is interesting that the anthropic principle refers exclusively to *Homo sapiens* and the purpose of our being here. But the dinosaurs, who held the Earth for 160 million years, had a better case to be considered lords of creation. Unaware of the cosmic accident waiting to remove them, the dinosaurs would have had good reason to invent a 'Reptilomorphic Principle' to explain why they had 'ruled the Earth' for so long. The beetles, all three million species, without counting in their multitude of ancestors, have an even stronger case. And what about the billions of foraminifera in the oceans, or the trees or the rocks, none of which would exist if the physical constants were different?

We may live in a 'Designer Universe' but it wasn't designed for us. Whatever the ultimate significance of the large numbers coincidence, the binding energy of the carbon-12 nucleus, or the fundamental physical constants, they were not the cause of the extinction of the dinosaurs nor of the emergence of *Homo sapiens* on the plains of Africa. If the great extinction at the close of the Palaeozoic had been a little more efficient, it would have set the evolutionary clock back two or three billion years to the blue-green algal stage. It might have stayed there and we might never have evolved.

The anthropic principle looks like another desperate attempt to put *Homo sapiens* comfortably back on centre stage, a view compatible with the authors of the Book of Genesis, the Koran and various other religious texts. However, as I mention a little later, a retreat into medieval modes of thought on our over-crowded planet would bring catastrophe.

The antidote to the anthropic principle: chance events?

In the previous section, I discussed the physical constants that look as though they were designed for *Homo sapiens* to arise. Here, in contrast, it is worth listing a few of the major chance events in the physical world that have directly affected the origin and evolution of life and our existence on Earth. I have mentioned others earlier throughout the text.

The list is impressive. I begin with the size of the fragment that broke away from the molecular cloud. If the fragment had been bigger, or spinning more rapidly, it would have spun itself out to a dumbbell and made two stars. Then it took fine timing to form Jupiter before the gas was all swept away. Without the shield of that giant planet, we would suffer a continual bombardment of comets. If another Jupiter had formed, we might have a system with only two planets, one closely orbiting the Sun every few days and the other far away in a wildly eccentric orbit. The accumulation of a few less planetesimals and comets would have made all the difference. The water on the Earth was accidentally contributed by comets but it makes plate tectonics possible. If the Earth was a little smaller or drier, the basaltic lavas could not recycle back into the mantle. Without water, we would have no granite, no continents to stand on, few ore deposits and no advanced technology. The continents enabled the final stages of land-based evolution to proceed above water and so enable this narrative to be written. The barren basaltic plains common throughout the other solid planets are less inviting.

The rotation rate of the Earth, which we take for granted, is a probable consequence of the great collision that formed the Moon. Without that accident, the Earth might have resembled Venus, and rotated slowly backwards. The tilt of the Earth that provides the seasons, celebrated by so many composers and painters, is the result of this same accident, which also removed any thick primitive atmosphere.

But perhaps the most dramatic accident of all was the massive dinosaur-killing impact that closed the Age of Reptiles. It seems certain that if the asteroid had missed, the descendants of the dinosaurs would now dominate the planet. Our ancestors would never have walked on the plains of Africa and I would not be sitting at a word processor writing this account. Albert Einstein (1879–1955) made a famous comment that 'God does not play dice'. However, the evidence for the importance of chance events indicates either that He does, or, as Laplace said in his famous reply to Napolean's question, we have no need for that hypothesis.

Is there a purpose?

Studies of the 'purpose' of life or of the universe and similar weighty matters are usually considered the province of philosophers. Today, however, their horizon seems to have become more limited than that of Kant, 250 years ago. He could make a significant advance in cosmology by his concept that the Milky Way was only one of a host of 'island universes'. Discussions over the meanings of words now seem to have become their major preoccupation. Certainly, their record in solving major questions pales beside the achievements of science.

While philosophers have sought the 'mind of God', and the ultimate purpose of the universe, the evidence that our existence was mostly a matter of chance has, in the presence of the Moon, literally been staring us in the face. This is ironic since for more than a century the solar system was relegated by the astronomers to an unimportant corner of the universe. It was so small that to a first approximation you could ignore it in thinking about important issues.

It is now the physical scientists, rather than the biologists, who have not accepted the significance of random events in the universe. Some behave like the man in the woodcut, trying to peer behind the curtain of the visible world in search of the clockmaker behind it. Physicists such as Paul Davies (b. 1946) want to see meaning, but in a solar system dominated by chance, any ultimate purpose can have no meaning. Paul Davies' philosophy can be summed up by his ringing statement that 'I cannot believe that our existence in this universe is a mere quirk of fate, an accident of history, an incidental blip in the great cosmic drama. This can be no trivial detail, no minor byproduct of mindless, purposeless forces. We are truly meant to be here'.[19]

Davies notes that some people such as Jacques Monod (1910–1966), the Nobel Prize winner, express a differing view. Monod has made the famous comment that 'nature is objective' and that 'man knows at last that he is alone in the unfeeling immensity of the universe, out of which he has emerged only by chance. Neither his destiny nor his duty have been written down'.[20]

Some people are dismayed by this evidence for our unique situation and seek refuge in the many varieties of mysticism that have appeared once again. One writer has commented that 'though science is stronger today than when Galileo knelt before the Inquisition, it remains a minority habit of mind and its future is very much in doubt. Blind belief rules the millennial universe, dark . . . as space itself'.[21] One is reminded of what has been called 'the failure of nerve' of the ancient world. Following the scientific advances of ancient Greece and the major progress at the Museum and Library at Alexandria, there followed a retreat into the comfort of myths. Much of the great library at Alexandria, which had found shelter in the Temple of Serapis from civil wars in the third century AD, was destroyed by a Christian mob in 391 AD. The remaining books were burnt in 646 AD by the Arab conquerors of the city. There followed a retreat from objective truth into medieval modes of thought as the Dark Ages began. Book burning seems to be a popular and widespread habit, judging from more recent examples that include Nazi Germany and the Red Guards of China.

Will faith once again smash science, as happened in the ancient world? This seductive pathway is not without its difficulties. As Harvey Brooks (b. 1915) has remarked 'if the modern era has created social and cultural conditions in which the enterprise of science is no longer viable, it has sown the seeds of its own disintegration and decay, to be followed by the disappearance of a large fraction of the world's present population and a decline in the material conditions of human life. It is a mere detail whether this will come about first through some ecological disaster, through the decay and demoralisation of the technological structure, or through a military holocaust'.[22]

However, it seems to me to be better to stand up and face the objective evidence for what it is, rather than behaving like the mythical ostrich and burying one's head in the sand. The knowledge that we are probably alone in the universe, that conscious intelligence has arisen accidentally, and we are its only keeper, should stimulate us to behave more responsibly. The shocking behaviour of many members of our species, displaying traits once useful for survival in primitive environments, and abetted by primitive tribal folklore and religious beliefs, forms an extraordinary contrast to this view.

The message of this book is clear and unequivocal: so many chance events have happened in the development of the solar system that any original purpose, if it existed, has been lost. Superimposed on these chance events from the physical world are those of biological evolution, which has managed to produce one highly intelligent species out of tens of billion attempts over the past four billion years.

The unique nature of the solar system

On the purpose of this inquiry

Why should we look into the nature, origin and history of the solar system? What is the point of describing all the infinite detail of the planets, satellites, rings, comets and asteroids that make up such a fascinating array? Is it merely to 'peep into nature's wonderland'? Or is the purpose to arrive at some conclusion about our present place in the universe and to try to understand how we have arrived at it? As Charles Darwin remarked 'If a person should ask my advice before undertaking a long voyage, my answer would depend on his possessing a taste for some branch of knowledge which could be acquired. It is necessary to look forward to a harvest when some fruit will be reaped'.[23]

Otherwise, there seems little point in extensive travels either, as in the case of Darwin, around the globe, or in our case, through the solar system. Merely to gape at the marvellous sights provides little more than one way of passing the time, like a tourist uncertain whether he is in Brussels or Berlin.

The accidental nature of the solar system: is it unique?

A certain amount of confusion has arisen over the question of the existence of other planetary systems. This has been driven by the natural wish to find clones of our system, or of the Earth in particular, complete with intelligent inhabitants. A statistical argument is often employed to argue that life must be common in the universe. There are well over 500 billion galaxies each containing over 100 billion stars. What percentage of these are single stars with planetary systems? One would suppose that no matter what statistical limits were applied, one would expect to find planets similar to the Earth, on which life will have arisen, and on which eventually intelligent life will evolve and will discover the 21 cm hydrogen wavelength as a suitable carrier for interplanetary, interstellar or intergalactic communication. The fallacy in this sort of argument, as Richard Dawkins has pointed out, is the assumption that a clone of the Earth will

develop and that life will arise and evolve to high intelligence as it has
done here. It is clear from the evidence both from the development
of the solar system, and of the many chances in evolutionary development, that this argument is false. When one multiplies the chances
discussed here of building a planet like the Earth with the chances of
developing high intelligence on it (one in many billions), the odds
would make even the most hardened gambler flinch.

Other planetary systems exist, no doubt with earth-sized planets
orbiting within 'habitable' zones. But our experience with our own,
with its strong evidence of random or stochastic processes, indicates
that the details in other systems can be predicted to be different.
Hence the philosophical question turns not so much on the occurrence of planetary systems, but on whether the detail of our own
system is unique.

The astonishing diversity that is observed within our own system
results from the application of the basic laws of physics and chemistry, but there is no simple recipe from which we can construct the
present solar system from first principles, any more than one could
predict the existence of elephants from a basic understanding of molecular biology. Attempts to force the strange compositions of the
Moon and Mercury into the grand plan represent futile attempts to
produce a grand unified theory. The division into terrestrial and giant
planets, the wide variety of satellites, the existence of that unique
satellite of the Earth, the Moon, the asteroid belt and the many other
unusual and astonishing details are as unlikely to be repeated as is the
course of the evolution of life on this planet. Local accidents have
predominated over general theories, just as some overlooked detail of
the landscape may ruin the course of a battle that was planned
according to the best principles of military strategy.

Here I have pointed out the difficulties of constructing a twin of
the Earth. Even Venus, close enough, one might think, to the Earth
in size, mass and density, is wildly unsuitable for life as we understand
it. As we saw, that planet is related to the Earth, its apparent twin,
only in a manner reminiscent of the connection between Dr Jekyll
and Mr Hyde. The arrival of a few more or less planetesimals as the
planets formed has produced two totally distinct planets. A little
closer to, or farther from, the Sun either cooks or freezes the planet.
But even being at the 'right' distance may not help. Too thick or too

thin an atmosphere may produce either an inferno that would have impressed Dante, or a frozen wasteland that a penguin might quail at. So the possibility that a copy might exist of our solar system, or of the Earth with all its exquisite detail, is judged to be unlikely.

What do other planetary systems look like?

This used to be a hypothetical question. When considering this problem, in 1992, I wrote[24] that

> other planetary systems will differ in the sizes and numbers of planets. What combination of circumstances could reproduce in some other system the detail observed in our own solar system or lead to the assembly of and evolution of a clone of the Earth? Other planetary systems doubtless exist. That they would resemble our ours in any but the broadest detail is only a remote possibility

Later in the same book, I wondered

> would we see something like the Galilean satellite system of a few equal-sized planets, systems with one giant planet, or a single brown-dwarf companion comprising a failed binary-star system?

Finally, after contemplating the satellite systems around our giant planets, I concluded that

> no simple sequence of reproducible events has occurred in our solar system. Other planetary systems, if discovered, will be different in detail to our own. What their satellites might look like is only for bold spirits to predict.

Now we have some preliminary answers. Around 20 bodies have been discovered orbiting other stars. Do these new discoveries have parallels with Galileo's observation of the phases of Venus and of four satellites in orbit around Jupiter? His discovery effectively destroyed the Ptolemaic System and led to the acceptance of the ideas of Copernicus. Perhaps the differences between these 'new planets' and our familiar solar system may finally lead to the realisation that we are alone. However, raw hope will no doubt continue to insist that somewhere out there is our double.

The first question is: what are we looking at? Not the bodies themselves, which are lost in the glare from the star. They have been detected by their gravitational pull on the star, since both bodies obey newtonian mechanics by rotating around their common centre of mass. In our own system, Jupiter, only a thousandth of the mass of the Sun, causes a wobble in the rotation of the Sun of thirteen metres per second. So if there is a similar body around another star, we can, in principle, detect this movement.

The most successful method of detecting bodies in orbit around other stars relies on the well-known Doppler effect. As the star wobbles in its orbit, it approaches to and recedes from the Earth with velocities of a few metres per second. These small changes cause the spectral lines to shift their wavelength slightly. As the star moves toward the Earth, the lines move toward the blue or shorter wavelength part of the spectrum. As the star recedes from the Earth, the lines move a little toward the red or longer wavelengths, a minute example of the 'redshift' that we observe in the spectra of the far distant galaxies that tells us of the expansion of the universe.

To measure this tiny shift in the wavelength of the spectral lines toward the blue or red part of the spectrum, one needs a spectrograph with very high resolution. The best of these now in use can detect a movement of the star towards or away from the Earth of around three metres per second. This high precision has indeed resulted in detecting star wobbles. At present only bodies around the mass of Jupiter can be detected; earth-sized bodies cause much too small a variation in the rotation of the star. Some stars wobble about due to internal instabilities and such effects may mimic the presence of nearby planets. The first 'planet' to be discovered, in very close orbit around 51 Pegasi was indeed thought to be such a mythical beast, but it seems to be real enough.

The immediate reaction of the new discoveries, like the report of possible life on Mars, was to raise the hopes of finding intelligent extra-terrestrial life, the 'little green men' so beloved of science fiction writers and credulous people generally. Like most other new discoveries in science, these bodies have, however, raised more questions than answers.

What has been observed? The new planets seem to fall into two groups. The first group, now totalling eight confirmed around different stars, are bodies that range in mass from one half to ten times the mass of Jupiter. These are minimum estimates of their mass because it is not possible from our viewpoint to determine the plane in which they are orbiting the star.

The most startling fact is that these 'planets' are mostly much closer to their star than Mercury is to the Sun and that they have orbital periods of a few days (see Figure 4). Even Mercury, 58 million kilometres away, but close enough to our Sun to be baked by it, takes 88 days for one orbit. One of the new 'planets' is only seven million kilometres from its star Tau Bootis. This planet is four times more massive than Jupiter and spins around Tau Bootis in a little over three days. Most of the other new planets, referred to as 'Hot Jupiters' are similarly very close in to their stars and orbit around them with periods of a few days. A couple are out at about two AU, with orbital periods of about two years. Four of these new bodies have highly eccentric orbits, but the others have nearly circular orbits like our own planets. At least one 'planet' near Beta Cygni, is in orbit around one member of a double star system. All these properties were unexpected.

To be discovered at all, the new 'planets' must be massive, mostly more so than Jupiter. However, no current model allows for the formation of a gas giant so close to its star that its year is accomplished in only a few of our days. The early violent evolution of a star will drive the gas, water and volatile elements needed to form the planet far out in the nebula. The formation of a gas giant in our system depends on forming a core that can, once big enough, capture gas around it. In our system, this happens at the ice condensation point, out around five AU.

Various models have been suggested to overcome this paradox posed by the 'hot Jupiters'. Some suggest the trapping of gas around dry rocky cores close in to the star, but how the gas remains when the ice has gone is a bit of a mystery. The most realistic model suggests that the giant planets indeed formed just like Jupiter. Ice piled up at a snowline where the temperature got low enough. A massive core formed that then captured the fleeing gas and a gas giant grew in the nebula.

While there was still some gas left, the giant could migrate inwards. Such a minor tidal evolution has been suggested for our own giants, which now may be a little sunwards of their initial position.

When the nebula gas had gone, the giant planets were left stranded at various distances from the star, just like whales that ventured too close to the shore and were left high and dry by the tide going out. The bottom line is that our hard-won theories for forming the giant planets in our solar system are robust, but that tidal evolution may produce spacings of the giant planets that seem bizarre to us.

The arrival of a Jupiter-like giant in the region where our rocky planets formed would create the sort of mayhem that accompanied the capture of Triton and that devastated the inner satellites of Neptune. Like a new broom sweeping clean, the giant would collect any surviving rocky debris, toss it into the star, or into the outer reaches of the nebula. Some interesting planets or asteroids formed from this debris might result in the outer reaches of such a planetary system.

Another plausible study begins with the formation of three gas giants form out in the nebula. Like an unruly mob, these giants interact by tossing one member out of the system, throwing another out into a wildly eccentric orbit far from the star, and moving the survivor into a close, but stable embrace, orbiting the star with periods of a few days. It has been surmised that we are lucky that only two gas giants, Jupiter and Saturn formed in our own system.

A second group of a dozen or so new objects is of much bigger objects, coming within the range of brown dwarfs. One at least, Gliese 229B, is a genuine example of these long awaited, but elusive creatures. They range in mass from 17 to 60 Jupiter masses and are mostly in highly eccentric orbits. Again, with few exceptions they are very close to the parent star, well within the Earth–Sun distance. Probably they have formed directly by condensation from the nebula at the same time as the companion star. These new objects are not universal. More than half the young stars surveyed show no evidence of disks and so will not form 'planets'. Most of the older stars show no evidence of planets.

I mention only in passing the 'planets' that have been reported around pulsars. Pulsars are dense neutron stars that are the residue left from a supernova explosion. Three or four 'planets' have been claimed, but some at least are dubious. How these 'planets' arose is uncertain. They seem unlikely to have survived the explosion of the star and so perhaps have formed subsequently, perhaps condensing from material ejected during the catastrophe. The bodies are spaced at varying distances from the star. Some are well within the orbit of Mercury in our system; one is as distant from its star as Uranus is from our Sun. These unfortunate bodies are bathed in X-rays and gamma rays from their rapidly rotating parent. Clearly, they are not a good environment for life and so are of less interest here than the other new examples.

These new discoveries reinforce the message from our own system. Nothing resembling our solar system has been discovered. Clearly the conditions that existed to make our marvellous set of planets are not easily reproduced elsewhere. As we have seen throughout this book, no two planets in the solar system are alike. Even the 'twins', Earth and Venus are so different that they could come from different planetary systems. The 60-odd satellites are just that: odd characters that defy efforts to put them into pigeonholes. This should have demonstrated to all that alien planetary systems would be unlike our own. So it should have come as no surprise that when nature tried elsewhere to build planets, or make some sub-class of brown dwarfs, that the end result was different. We are left with the conclusion that attempts to find some general formulae for recreating the detail of the solar system, a problem to which I now turn, are likely to be on the wrong track.

Attempts to find a general theory to build planets

Stephen Brush, in his perceptive review of theories for the origin of the solar system noted that 'attempts to find a plausible naturalistic explanation of the origin of the solar system began about 350 years ago, but have not yet been quantitatively successful, making this one of the oldest unsolved problems in modern science'.[25] A basic question is whether one could produce the solar system 'from first principles', so that by feeding in the appropriate data for the initial starting conditions in the nebula into a large enough computer, the present solar system would appear at the end of the calculation.

In studying the natural world, it is difficult to avoid being overwhelmed by detail. Attempting to see the forest for the trees has always been a difficult exercise; the study of chemistry before the discovery of the Periodic Table provides a classic example of a bewildering set of data that was eventually revealed to possess an underlying physical order. The diversity of organisms in biological systems became understandable as resulting from the operation of a single universal law, Darwinian evolution, although the extreme complexity of organisms which have arisen have made any further generalisations difficult. To add further complexity, the course of evolution is driven by chance events, and hence is unpredictable.

In the solar system, a similar diversity results from the application of the basic laws of physics and chemistry. Hence one is unlikely to find any grand blueprint for constructing such systems from the study of one system; so much of the detail is due to the operation of random processes. Thus, attempts to find some uniform principles, analogous to the Periodic Table or Darwinian evolution, from which one might construct clones of our solar system, appear to be headed in the wrong direction.

Rather than constructing such grand unified theories, a whole variety of scientific questions that need to be addressed have appeared. Until very recently, the problem of the origin of the solar system has been treated as an intellectual puzzle. Workers took the small number of boundary conditions and attempted to find a complete unified solution. In this endeavour, it was customary to give a list of questions that need to be answered by any theory of solar system origin.

Usually a dozen or so observations, such as the concentration of mass in the Sun and angular momentum in the planets, the slow rotation of the Sun, the co-planar and prograde rotation of the planets, the Titius–Bode rule, the number of planets and satellites, the distinction between the terrestrial and giant planets, planetary obliquities, and so forth are listed as important problems to be explained. But if these observations are unique to our solar system, attempts to explain them cannot lead to a general theory of the formation of planetary systems.

Numerous answers have been provided over the past 300 years for such questions, all claiming to have solved the problem. Seeking such theories is a false goal, since we cannot identify one dominant process. This view is reinforced by the discoveries of the strange new 'planets'.

It is only recently that the existence of the solar system has been regarded as an ordinary scientific problem, in which workers investigate individual pieces of the problem. As this book has shown, we have moved into a more realistic and pragmatic view of the solar system in which we live. We are dealing with a system in which many random events occurred, with the end result that all planets and satellites are different.

Much of our difficulty in trying to understand the solar system stems from the fact that the Earth, with its complex and unique history and obscure cratering record, was not the best place from which to begin. The Moon also turned out to be a unique object. There it is, in plain sight, accessible to naked eye observation, the closest, but until recently one of the most enigmatic objects in the universe. All the time, it was telling us that chance events were common.

The end of the solar system

We may reasonably expect that the system will continue along its present way for the next four billion years or so. Although the system is 'chaotic' in a mathematical sense, this merely means that we cannot accurately predict the exact location of the Earth a few hundred million years in the future (or, for that matter, in the past). The Earth and the other planets, however, are likely to be in similar and nearly circular orbits and distances from the Sun for the next few billion years. This reassuring news is supported by evidence for stability in the geological record of the Earth. As we saw earlier, running water, which has been eroding rocks and depositing sediments, has been around for the past four billion years, so that the Earth has not wandered far from its present orbit.

The Sun, however, will eventually come to the end of its tether. Like a power station that has run out of fuel, it will shut down, but in a somewhat more dramatic fashion, although more slowly, at least in the initial stages. As the hydrogen in the core is consumed and the fusion to helium grinds to a halt, the Sun will start to collapse as the fire goes out. As the internal pressures build up due to the collapse, it becomes hot enough for a second cycle of nuclear fusion, involving helium, to begin. The Sun will swell up to become a red giant, expanding out to engulf Mercury within a few million years. However, it will have lost perhaps a quarter of its mass in the process, so that Venus and the Earth, and the other planets, may retreat from the dying giant as its gravitational clutch weakens. The Sun will shrink in volume again as the helium fuel is exhausted and the fire dies down once again. Again, the increasing pressure will reignite the nuclear furnace. The Sun will balloon out to the present orbit of Venus in a second true red giant stage.

By this time, it will have lost a third of its mass. Venus might retreat out of reach of the weaker Sun. Conditions on the Earth will not easily be imagined. The Sun will be ten times larger and over 2000 times more luminous, and will occupy a large fraction of the sky. H. G. Wells (1866–1946), a hundred years ago in his book, *The Time Machine* (1895), imagined that far in the future his time traveller, who had come to rest on a desolate beach, saw this scene where the ocean was 'all bloody under the eternal sunset' and 'the huge red-hot dome of the Sun had come to obscure nearly a tenth part of the darkling heavens'.

Further catastrophes will follow this red giant stage. The Sun will shed most of its mass. It will finish up as a white dwarf, about the size of the Earth, as the inexorable crushing force of gravity finally wins its ten billion year battle against the expansionist forces of heat. Not much will be left of the solar system after these violent episodes. Our familiar inner planets, Venus, Earth and Mars, with their marvellous diversity, will all have been melted or swallowed up by the Sun. There will be no more brilliant morning or evening stars, dewy dawns, lazy summer days or snowy winters, red sunsets from great volcanic eruptions, harvest moons or poetry.

> The cloud capped towers, the gorgeous palaces,
> The solemn temples, the great globe itself,
> Yea, all which it inherit, shall dissolve,
> And like this insubstantial pageant faded,
> Leave not a rack behind[26]

Giant Jupiter and its satellites, splendid Saturn and its rings, green Uranus and blue Neptune, all will be cooked to a cinder as the bloated red giant sheds it mass out into space, like a crazy millionaire throwing his wealth away. This will provide the material, enriched in carbon, oxygen and a few heavier elements, products of the nuclear furnace in our Sun, to form new stars. The elements in us were formed on one star and will end up in another, recycling on the grandest scale.

What is remarkable is that the physical sciences can predict, with great certainty, that these events will occur four or five billion years in the future. Economics, along with most other human affairs, is in much worse shape. The rise and fall of stock markets on a monthly scale is due to much more dimly understood factors than the history of the Sun. And what can one say about the political sciences that failed to predict the collapse of the USSR until that significant historical event was upon us.

The universe will continue along its way, heedless of the trivial and commonplace disturbance of a dying star and of the one-time existence of the solar system. Five billion years in the future, the universe will present its familiar appearance, little changed from the birth of the solar system, ten billion years earlier. Iron, oxygen, carbon, gold and silver and the array of the other elements will be a little more abundant in the clouds of gas and dust in the spiral arms of the galaxy. The constellations, seen from our perspective, will have long since rearranged themselves, a few tens of thousands of years only sufficing for the relative movements of the nearby stars to change them to new shapes. The Big Dipper, The Pleiades, Orion, the Hunter and his companion, the Great Dog, all will have long vanished.

The atoms of the word processor on which this is being written, and of the operator himself, may find themselves, after wandering in the cold reaches of space, caught up in a molecular cloud, and become engulfed in a new star, or perhaps become part of another planetary system.

Notes

The following are the sources of the quotations in the text and various notes on topics mentioned as asides to the main topic of the book.

1 Setting the stage

1 Laplace, P. S. (1796) *The System of the World*, Vol. 1, Book V (English translation by J. Pond, 1809, R. Phillips, London), p. 293.

2 Boorstin, Daniel J. (1985) *The Discoverers*, Vintage Books, New York, p. 296.

3 Jaki, S. L. (1978) *Planets and Planetarians*, Wiley, New York, p. 26.

4 Dick, O. L. (1958) *Aubrey's Brief Lives*, Secker and Warburg, London, p. 94.

5 Brush, S. L. (1996) *A History of Modern Planetary Physics*, Vol. 1, Cambridge University Press; p. 20 gives an account of the celebrated exchange between Laplace and Napoleon.

6 Readers interested in the phenomenon of 'Unidentified Flying Objects' may make a sober start by referring to the historical account on 'The UFO controversy and the extraterrestrial hypothesis' in Dick, S. J. (1996) *The Biological Universe*, Cambridge University Press, pp. 267–307.

7 Shakespeare, W. (1599) *As You Like It*, Act III Scene 2.

8 Jonson, B. (1610) *The Alchemist*, Act II, Scene 3.

9 Kelvin, Lord (Thomson, W.) (1891) *On the Origin of the Sun's Heat, Popular Lectures and Addresses*, Vol. 1 Macmillan, London, 2nd edition, pp. 421–422.

10 The story of the unfortunate elephant, although often told, is not mentioned in the three standard works on the investment of the city: Goure L. (1962) *The Siege of Leningrad*, Stanford University Press, 363 pp.; Pavlov D. V. (1965) *Leningrad 1941*, University of Chicago Press, 186 pp.; Salisbury H. E. (1969) *The 900 Days: The Siege of Leningrad*, Harper and Row, New York, 635 pp.

11 An authoritative and very entertaining account of the fall and subsequent history of this meteorite is given by Marvin, U. B. (1992) *Meteoritics*, Vol. 27, pp. 28–72.

2 The giants

1 Cooper, Henry S. F. (1990) *The New Yorker*, June 18, p. 73.
2 Wetherill, G. W. (1989) in *The Formation and Evolution of Planetary Systems* (editors H. A. Weaver and L. Danly), Cambridge University Press, p. 27.
3 J. A. Burns (1986) in *Satellites* (editors J. A. Burns and M. S. Matthews), University of Arizona Press, p. 17.

3 Escapees and survivors

1 Shakespeare, W. (1599) *Julius Caesar*, Act II Scene 2.
2 Laplace, P. S. (1809) *The System of the World*, Vol. 1, Book I (English translation by J. Pond, 1809 R. Phillips, London), p. 97.
3 *The Bible*, King James Version, Oxford University Press, Psalm 90, verse 4.
4 The quotation by Edmond Halley is given in Fernie, J. D. (1985) *American Scientist*, Vol. 73, p. 471.
5 Cooper, Henry S. F. (1990) *The New Yorker*, June 18, p. 84.
6 Laplace, P. S. (1809) *The System of the World*, Vol. 1, Book I (English translation by J. Pond, 1809, R. Phillips, London), pp. 88–89.
7 O. Mitchell (1869), quoted by Brush, S. G. (1996) *A History of Modern Planetary Physics*, Vol 1, Cambridge University Press, p. 93.
8 Greenberg, R. and Brahic, A. (1984), in *Rings*, University of Arizona Press, p. 4.

4 The twins

1 Playfair, John (1802) *Illustrations of the Huttonian Theory of the Earth*. Facsimile reprint, G. W. White, University of Illinois Press, 1956, xiii-xiv.
2 Boorstin, Daniel J. (1985) *The Discoverers*, Vintage Books, New York, p. 86.
3 The best account of the geological history of the Earth is given by Cloud, Preston (1988) *Oasis in Space*, Norton, New York.
4 McSween, H. Y. Jr. (1989) *American Scientist*, Vol. 77, p. 146.
5 Shakespeare, W. (1606) *Macbeth*, Act I Scene 3.
6 e.g. Lovelock, J. (1988) *The Ages of Gaia*. Oxford University Press, p. 19.
7 e.g. Lovelock, J. (1988) *The Ages of Gaia*. Oxford University Press, p. 111.

5 Two special cases

1 Laplace, P. S. (1796) *The System of the World*, Vol. 1, Book IV (English translation by J. Pond, 1809, R. Phillips, London), p. 94.
2 The Rosetta Stone was discovered in 1799 during the French invasion led by Napoleon, near Rosetta, now Rashid, in Egypt. It is a tablet of black basalt, on which are listed benefactions by Ptolemy V Epiphanes (205–180 BC). These were inscribed by the priests at Memphis in two Egyptian scripts (hieroglyphs and demotic, script related to hieroglyphs) and also in Greek. This discovery enabled the ancient Egyptian picture language (hieroglyphs) to be translated, a task accomplished mainly by J.F. Champollion by 1822.
3 Dawkins, R. (1987) *The Blind Watchmaker*, Norton, New York, p. 6.
4 Huygens, C. (1698) *The Celestial Worlds Discovered*, T. Childe, London, p. 131.
5 Breen, J. (1854) *Planetary Worlds*, Robert Hardwicke, London, 250 pp.
6 Laplace, P. S. (1809) *The System of the World*, Vol. 1, Book IV (English translation by J. Pond, 1809, R. Phillips, London), p. 94.
7 The Gordian Knot was reputedly tied by King Gordias (father of Midas) in the town of Gordian, capital of Phrygia (in modern west-central Turkey), about 700 BC. By all accounts, it was a knot of great thickness and complexity, which secured the yoke to the harnessing pole or shaft of a two horse chariot. King Gordias prophesied that whoever could untie the knot would conquer Asia, which then essentially comprised the known world. All attempts failed until Alexander, The Great (356–323 BC), on passing through Gordian in 333 BC on his expedition against the Persians, solved the problem by cutting through the Gordian Knot with his sword. Alexander's subsequent conquest of Asia was held to have fulfilled the prophesy. The tale has survived to illustrate one manner of solving apparently intractable problems by drastic means.
8 Scott, R. F. (1977) *Earth Science Reviews*, Vol. 13, p. 379.

6 Causes and effects

1 Melosh, H. J. (1989) *Impact Cratering: A Geological Process*, Oxford University Press, p. 131.
2 Ralling, C. (1982) *The Voyage of Charles Darwin* (His Autobiographical Writings), Aerial Books, London p. 73.
3 Erwin, D. H. (1994) *Nature*, Vol. 367, p. 231.

4 Tremaine, S. (1986) in *The Galaxy and the Solar System* (editors R. Smoluchowski *et al.*) Arizona University Press, p. 413.

5 Crick, F. (1994) *The Astonishing Hypothesis: The Scientific Search for the Soul*, Simon and Schuster, New York, 317 pp.

6 A very useful discussion on the origin of life is given by C. F. Chyba and G. D. McDonald (1995) The origin of life in the solar system, in *Annual Reviews of Earth and Planetary Science*, Vol. 23, pp. 215–249. See also Eigen, M. (1992) *Steps toward Life*, Oxford University Press. This is the best statement of the problem since Jacques Monod's *Chance and Necessity* (1974). See note 20.

7 An account of the possible development of life under these circumstances is given by Russell M. J. and Hall, A. J. (1997) *Journal of the Geological Society of London*, Vol. 154, pp. 377–402 and by Huber, C. and Wächtershäuser, G. (1997) in *Science*, Vol. 276, pp. 245–247.

8 Cloud, Preston (1988) *Oasis in Space*, Norton, New York, p. 267.

9 Mayr, E. (1994) in *Perspectives in Biology and Medicine*, pp. 150–154; see also *The Planetary Report*, Vol. 16 (3), p. 6 (1996). An authoritative account of the debate over extraterrestrial life is given by Dick, S. J. (1996) *The Biological Universe*, Cambridge University Press, 578 pp.

10 Pollard, W. G. (1979) The prevalence of Earth-like planets. *American Scientist*, Vol. 67, p. 654.

11 The best discussion can be found in *The Search for Life on Mars*, by H. S. F. Cooper (1979), Holt, Rinehart and Winston, New York, 254 pp. See also note 9. The cartoon is from *The New Yorker* (1977), Vol. 53, No. 1, p. 27.

12 Ralling, C. (1982) *The Voyage of Charles Darwin* (His Autobiographical Writings), Aerial Books, London p. 139.

13 Dawkins, R. (1987) *The Blind Watchmaker*, Norton, London, pp. 5–6.

14 Gould, S. J. (1991) *Wonderful Life: The Burgess Shale and the Nature of History*, Penguin, London, 347 pp.

15 Dawkins, R. (1987) *The Blind Watchmaker*, Norton, London, p. 9. An extended discussion on the evolution of the eye and much else, is given by Richard Dawkins (1996) in Chapter 5 of *Climbing Mount Improbable*, Viking Penguin Books, London.

16 Shakespeare, W. (1611–1612) *The Tempest*, Act I Scene 2.

17 The anthropic principle is discussed at length in *The Anthropic Cosmological Principle* by J. D. Barrow and F. J. Tipler (1986) Oxford University Press, 706 pp.

18 A value of 3.0 for π occurs in the description of the building of the Temple of Solomon in Jerusalem in I Kings Chapter 7, verse 23 and in II Chronicles Chapter 4, verse 2. The often repeated story that a state legislature (variously Illinois, Indiana or Massachusetts) in America was sufficiently impressed by the biblical value of π that they attempted in the nineteenth century to pass a

law stating that the value of π was 3.0 appears to be without foundation (see Beckmann, P. (1971) *A History of Pi (π)*. Golem Press, Boulder, CO, p. 170).

19 Davies, P. C. W. (1992) *The Mind of God*, Simon and Schuster, New York, p. 232.

20 Monod, J. (1974) *Chance and Necessity*, Collins Fontana, London. p. 154, 167.

21 Timothy Ferris, *The New Yorker*, April 14, 1997, p. 31

22 Brooks, H. (1971) *Science*, vol. 174, p. 21.

23 Ralling, C. (1982) *The Voyage of Charles Darwin* (His Autobiographical Writings). Aerial Books, London p. 130.

24 Taylor, S. R. (1992) *Solar System Evolution: A New Perspective*, Cambridge University Press, pp. xi, 38, 251.

25 Brush, S. G. (1996) *A History of Modern Planetary Physics*, Vol 3. Cambridge University Press, p. 91.

26 Shakespeare, W. (1611–1612) *The Tempest*, Act IV Scene 1.

Index